Advanced Electrino Physics
Draft 2

ELECTRON   PION

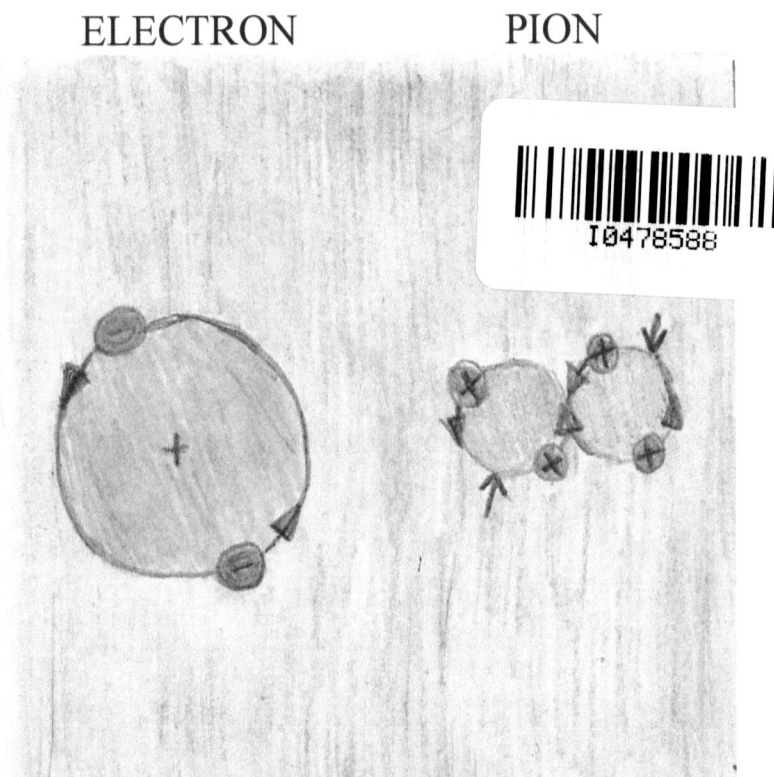

Figure 1. An electron is composed of two semions orbiting about each other.

Figure 2. A net zero spin pion is composed of two orbiting pairs of orbiting quartons.

by Gordon L. Ziegler and Iris Irene Koch

Advanced Electrino Physics
Draft 2

by Gordon L. Ziegler and Iris Irene Koch

Advanced Electrino Physics Draft 2

Copyright

ISBN: Softcover   978-1-4931-4591-1
              eBook   978-1-4931-4592-8

This book was printed in the United States of America.

Rev. date: 05/01/2014

To order additional copies of this book, contact:
Xlibris LLC
1-888-795-4274
www.Xlibris.com
Orders@Xlibris.com
543853

Authors
Gordon L. Ziegler and Iris Irene Koch
PO Box 1162
Olympia, WA 98507-1162 USA
e-mail: *ben_ent100@msn.com*

PREFACE TO DRAFT 2

Current models of physics, advanced as they are, cannot calculate the masses of elementary particles from first principles. The Electrino Fusion Model of Elementary Particles, however, is now at the stage when such calculations can be made. This book will calculate the $g/2$ factors of all the single particles used in particle structures up to state 4 or 5 in Chapter 5. These are necessary to calculate the masses of all the fundamental whole particles up to state 5. Except for 22 elementary particles calculated in this volume, the actual calculation of the masses of known particles shall be done in the next volume in this series, *Predicting the Masses,* by the authors.

Advanced Electrino Physics Draft 2 here also presents the theory of how to fuse the quartons in pions to the semions in electrons or positrons. This light is given to complete the knowledge base of the researcher. But this process is not pursued by the authors in that they think it is more hazardous than semion fusion, which could lead to many fatalities if it were tested.

An additional difference is discussed in this volume between total spin and observable spin. This intelligence should be useful to the researcher from now on in the calculations in particle physics.

In the previous volumes, *Electrino Physics* and *Advanced Electrino Physics,* different versions of CODATA database and Particle Data Group database were employed ranging from 1998 to 2006 dates. In Draft 2 of these volumes, the CODATA data is updated to the most recent 2010 version, and the Particle Data Group data is updated to the most recent 2013 and 2014 versions. The difference between the measured Fine Structure Constant α in the 2010 version and the 2006 version is in the right direction to harmonize the measured and calculated $g/2$

factors for the electron and muon, but it is not enough yet to harmonize them completely by this means. The inaccuracies of the data are because meters, kilograms, seconds, and Coulombs are not defined precisely enough. In natural units the data is already exact.

CONTENTS

CONTENTS

Chapter 1

FUSION OF QUARTONS

This book, *Advanced Electrino Physics*, continues the presentation of the material started in the book, *Electrino Physics*, but beyond the scope of that book. The first subject in that category is the fusion of quartons. While the author long believed that ionized quartons could fuse to semions, he did not know how to fuse bound quartons (as in pions, kaons, and D-ons), because those are zero spin particles and are bosons. They can go right through each other without colliding. Therefore they were not like positrons and electrons, of which the author theorized how to fuse the semions or anti-semions in them. The boson character of quarton systems presented a long insurmountable barrier in the author's mind to theorizing how to fuse quartons in bound systems.

The second crack in that barrier occurred as a result of conversations with Iris Koch, the author's sister, on the structure of pions. While ground state pions apparently have in them two quartons orbiting one way and two quartons orbiting the same way, and the two pairs of quartons orbiting the opposite way, we discussed a possible four-body state of the orbits being at right angles to each other, or other relative angles. We realized that pions, kaons, and D-ons could change relative orbit angles freely depending on energy state conditions.

The first crack in that barrier occurred years ago as a result of the author observing boson gravitons being ripped apart as the magnetic field in electrons tried to realign the orbits of the positrons and electrons in the gravitons, as seen in balancing decay schemes in chonomic equations for leptons.

On March 18, 2007, those ideas were put together in the author's mind. He first thought of realigning the quarton orbits in pions through a strong magnetic field. Seconds later he was impressed of doing it by means of a fast electron. He then looked in the 80th Edition of the *CRC Handbook of Chemistry and Physics*[1] to see if he could see any annihilation γγ decay modes for pions. He could see none. He could see none also for the D particle. But he found such a decay mode for kaons: $K^+ \rightarrow \pi^+\gamma\gamma$. He was at first surprised to see the π^+ particle in the decay products. But then he immediately saw that that π^+ did not come directly from the K^+ particle. It was the tag along left over particle with an electron in an unobserved electron neutrino colliding with the kaon. The electron was the active particle in the neutrino collision. First it realigned the orbits of the quartons in the kaon to make them fuse to two anti-semions in a positron. Second it annihilated with the resultant positron, leaving behind the left over pion. The pion in the reaction is a sure signature of the neutrino, proving this reaction did not occur naturally by other means.

Almost immediately the author thought of doing this artificially using high speed electrons rather than neutrinos. The electrons would be more controllable than neutrinos. In that case, $K^+ + e^- \rightarrow \gamma\gamma$. There would be no π^+ in this reaction. Using the skills taught in *Electrino Physics*[2] Chapter 10, Appendix A, and Appendix B, we can write chonomic equations for these two reactions:

Observed $K^+ \rightarrow \pi^+\gamma\gamma$.

An unobserved high speed electron neutrino, with ½ℏ positional-kinetic angular momentum, collides with a K⁺. The magnetic field of the electron in the neutrino penetrates the K⁺, which is made up of two orbiting pairs of quartons, and rotates the axes of the orbiting quartons to the same direction. The four quartons then fuse to two semions of a positron, which annihilates with the neutrino electron, leaving the neutrino π^+ and two oppositely directed energized pre-existing annihilaton photons (gammas).

Theorized K⁺ + e⁻ → γγ.

```
              collision,              annihi-
  K⁺     e⁻    fusion    γ       γ    lation     γ       γ
 493...  0.51...  q.p.1   0       0    q.p.2      0       0
  |o       |        |     •|•     •|•   ••|••      •|•      •|•
  |    +    |   →    |  +  |   +   |   →  |    →   |    +   |
  |        -|       -|+    |       |      |        |        |

 0|0     0|-½      0|0   -1|-1    1|1   1-1|0     1|1     -1|-1
 0|0     ½|0       0|0    0|-1    0|1    0|0      0|1      0|-1
                        |_____|
                          unobserved
                          particles
```

An axial spin, high speed electron, with ½ℏ positional-kinetic angular momentum, collides with a positive kaon. The magnetic field of the electron penetrates the kaon, which is made up of two orbiting pairs of quartons, and rotates the axes of the orbiting quartons to the same direction. The four quartons then fuse to two semions of a positron, which annihilates with the accelerated electron, leaving the oppositely directed energized pre-existing annihilation photons (gammas).

The first reaction is already observed and reported. This gives us confidence that the similar second reaction may take place as theorized. The reaction might be spin orientation sensitive. So the experimenter should not give up at a first failure. The end product is worth it. It could help relieve our energy crisis.

By targeting anti-kaons with positrons, similarly to above, fusions and annihilations would occur. By targeting anti-kaons with electrons, stable electrons would be produced by anti-quarton fusion. No annihilation would occur. Also, in either case, the second law of thermodynamics would be reversed by these reactions, if the repetition rate were kept down. This could be an alternate method to that reported in Chapter 16 of *Electrino Physics*[3] for reversing the order to disorder arrow in the second law of thermodynamics. This might be done at an existing large accelerator laboratory. This may infuse new interest in the accelerators.

---

[1]SUMMARY TABLES OF PARTICLE PROPERTIES, January 1, 1998, Particle Data Group, as quoted by *CRC Handbook of Chemistry and Physics, 80th Edition*, David R. Lide, Editor-in Chief (Boca Raton: CRC Press, 1999), pp. **11**-1 to **11**-49.

[2]Gordon L. Ziegler, *Electrino Physics* (P.O. Box 1162, Olympia, WA 98507-1162 USA; e-mail: ben_ent100@msn.com: Book available for downloading free at http://benevolententerprises.org Book List. To order copies of this book, contact: Xlibris LLC, 1-888-795-4274, www.Xlibris.com, Orders@Xlibris.com.

[3]*Ibid.*

Problem Set 1

1. What characteristic of quarton systems presented a long insurmountable barrier in the author's mind to theorizing how to fuse quartons in bound systems?

2. What is the second crack in that barrier that occurred?

3. What is the first crack in that barrier that occurred years ago?

4. How were those ideas put together in the author's mind?

5. Did the author find an annihilation natural decay for pions?

6. Why is that a tremendous blessing?

7. What quarton particle did the author find that did have a listed annihilation natural decay?

8. What unobserved particle is involved in the fusion of that particle.

9. What additional unobserved particles are involved in the annihilation of that resultant particle?

10. What value of positional-kinetic angular momentum occurs in the fusion of this particle that never occurred in any lepton decay in *Electrino Physics* Appendix A?

11. What values of positional-kinetic angular momentum occurred in lepton decay in *Electrino Physics* Appendix A?

12. In your opinion, is the value of positional-kinetic angular momentum in question 10 proper? Is it possible?

Why? Would any lower positive value of positional-kinetic angular momentum, other than zero, ever be observable?

13. The $K^+ \rightarrow \pi^+ \gamma \gamma$ decay scheme occurs naturally. In this reaction, does the π^+ come from the K^+ directly, as by knocking the K^+ echon down to the lower energy state? Where does the π^+ in the above reaction come from?

14. What decay scheme do we hope can be induced artificially?

15. What would be the only observed products of such a reaction?

16. Why, even if the reaction worked, might it not work the first time it was tried?

17. How could stable electrons be produced by quarton fusion?

18. Would such a reaction reverse the order to order disorder arrow in the second law of thermodynamics?

19. Would this require a high repetition rate with a high beam current of anti-kaons, or a low repetition rate with a few anti-kaons?

20. What would undoubtedly occur in the public perception of accelerators if this latter reaction occurred?

Chapter 2

HARMONIZING PARTICLE SPINS

In early editions of *Electrino Physics*, Chapter 6, in Postulate 8 and later derivations for the electron in Section IIIA, the calculated spin of the electron is $\hbar$, whereas the traditional spin of the electron is $\hbar/2$. The explanation in the model is that $\hbar$ is the total spin of the electron, whereas $\hbar/2$ is the observable spin of the electron.

The difference between total spin and observable spin is due to the structure of the particles. All particles are mass singularities. However much spin there may be inside a mass singularity, the only amount of spin that is observable from a mass singularity is due to the fracton on the side from which the spin is measured traveling at the speed of light at the event horizon of the mass singularity. The observer can observe no velocity faster than the speed of light, so he can observe only the spin that can be communicated at the event horizon of the singularity. The actual velocity of the fracton electrinos in the singularity may greatly exceed the speed of light (see the next chapter), but the greatest spin sense observable from the singularity can only be communicated at the event horizon of the black hole. Also, the observer cannot see *through* the mass singularity to see the fracton electrino on the opposite side of the singularity. That electrino contributes to the total spin of the particle, but not to the observable spin of the particle. For semion systems, the total spin may range from $\hbar$ to ∞ (see next chapter), but the observable spin is only always $\hbar/2$, because only one semion is observable at a time, with an effective mass of half the particle, at the radius r, and traveling at the maximum of the speed of light.

In the case of electrons, the two semions actually travel at the speed of light in their orbit. But their system is a mass singularity. The observer can only observe the effect of one semion. Half the mass of the electron times the radius of the electron times the speed of light c equals $\hbar/2$. That is the observable spin of the electron. The total spin of the electron takes into consideration the contributions of both semions. We have, then, two times half the mass of the electron, or the mass of the electron times the radius of the electron times the speed of light c equals $\hbar$.

For a muon, the semions actually orbit in the mass singularity at 11.7062 c (see next chapter, Section I). This makes the total spin for the muon equal to $11.7062\,\hbar$. But the observable spin of the muon (and all simple semion systems) again, as explained above, is just $\hbar/2$. No matter what the energy state of the semion system, the observable spin is the same. This is the angular momentum that can be conveyed and transferred in particle collisions.

The spin dynamics of mass singularites are strange but simple. Mastering this bit of science will greatly help in the study of the rest of this book.

Problem Set 2

1. What is the total spin of the tauon?

2. What is the observable spin of the tauon?

Chapter 3

PREDICTED MASSES OF CHARGED LEPTONS

So as this book may prepare the way for the calculation of the masses of every known particle, and may predict so far undetected particles, this book will here repeat, under a new title, Chapter 21 of *Electrino Physics*.

A. Introduction

In early chapters of *Electrino Physics*, the idea was expressed that electrons, muons, and tauons were just energy states of one particle system—and similarly for pions, Kaons, and D-ons as well as other particle sets. The author thought to solve for the various energy states like Niels Bohr solved for the energy states in hydrogen in 1913.[1] Bohr's calculational framework has been very helpful as a guide to the author in solving for the velocities, radii, and masses of particles in elevated states. This chapter will calculate these things. However there are many significant differences in the calculations. These will be pointed out.

B. The Bohr Atom

Bohr's results followed from algebraic derivations from a few postulates:

"1. The electrons move in orbits restricted by the requirement that the angular momentum be an integral multiple of h/2π, that is, for circular orbits of radius r, the electron velocity v is restricted by

$$mvr = \frac{nh}{2\pi} \qquad (3\text{-}1)$$

and furthermore the electrons in these orbits do not radiate in spite of their acceleration. They were said to be in stationary states."[2]

"2. Electrons can make discontinuous transitions from one allowed orbit to another, and the change in energy, E-E' will appear as radiation with frequency

$$v = \frac{E - E'}{h} \qquad (3\text{-}2)$$

An atom may absorb radiation by having its electrons make a transition to a higher energy orbit."[3]

3. Bohr obtained another relevant calculational equation simply by balancing the Coulomb electric force against the centrifugal force.

$$\frac{kq_e^2}{r^2} = \frac{m_e v^2}{r}, \qquad (3\text{-}3)$$

where $k = 1/(4\pi\varepsilon_0)$, and q_e is the charge of the electron.[4]

4. "The energy of an electron in an orbit is the sum of its kinetic and potential energies:

$$E = E_{kinetic} + E_{potential} \qquad (3\text{-}4)$$

$$= \frac{1}{2} m_e v^2 - \frac{kq_e^2}{r} ."[5] \qquad (3\text{-}5)$$

C. Electron Energy Levels in Hydrogen

Performing simple algebraic operations, Bohr was able to solve for the orbital velocity v, the radius r, and the energy E.

"To begin, multiply both sides of Eq (3-3) by r to see

$$\frac{kq_e^2}{r} = m_e v^2. \qquad (3\text{-}6)$$

The term on the left hand side is the potential energy. So the equation for the energy becomes

$$E = \frac{1}{2}m_e v^2 - \frac{kq_e^2}{r} = -\frac{1}{2}m_e v^2. \qquad (3\text{-}7)$$

Now we just need to figure out what the velocity, v is equal to, so solve Eq (3-1) for r,

$$r = \frac{n\hbar}{m_e v}. \qquad (3\text{-}8)$$

Plug this into Eq (3-6),

$$kq_e^2 \frac{m_e v}{n\hbar} = m_e v^2. \qquad (3\text{-}9)$$

Then divide both sides by m$_e$v to see

$$\frac{kq_e^2}{n\hbar} = v. \qquad (3\text{-}10)$$

Now we can put in this value for v into the equation for energy, and then also plug in the values for k and $\hbar$, and we'll obtain the energy of the different levels of hydrogen:

$$E_n = \frac{-1}{2}m_e \left(\frac{kq_e^2}{n\hbar} \right)^2 \tag{3-11}$$

$$= \frac{-1}{2}m_e \left(\frac{1}{4\pi\varepsilon_0}q_e^2 \frac{2\pi}{nh} \right)^2 \tag{3-12}$$

$$= \frac{-m_e q_e^4}{8h^2 \varepsilon_0^2} \frac{1}{n^2}. \tag{3-13}$$

Or, after substituting values for the constants,

$$E_n = \left(-13.6\ eV \right)\frac{1}{n^2}. \tag{3-14}$$

Thus, the lowest energy level of hydrogen (n = 1) is about -13.6 eV. The next energy level (n = 2) is -3.4 eV. The third (n = 3) is -1.51 eV, and so on. Note that these energies are less than zero, meaning that the electron is in a bound state with the proton. Positive energy states correspond to the ionized atom where the electron is no longer bound, but is in a scattering state."[6]

D. Three More Quantum Numbers

"Bohr had pictured the electron orbits around the atomic center as being perfectly circular, but this was too simple. There are very few perfect circles in nature, and orbits in atoms are no exception.

"Later, in 1916, the German physicist Arnold Sommerfeld refined Bohr's 'easy' picture with one a bit more complex. In this modified view the electron orbits were not circular, but elliptical. But there are many kinds of ellipses possible (certainly more than one), and this changed the calculations in subtle ways, as each ellipse has a slightly different angular momentum. To take account of the possibility of elliptical orbits, Sommerfeld introduced another number; the **orbital quantum number** (sometimes called the "angular momentum quantum number"), which usually had the symbol "L" [or "l"]. . . .

"Like the principal quantum number, the orbital quantum number can have values of 0, 1, 2, 3, 4, etc., but only up to a whole number value of one **less** than the electron's principal quantum number (i.e. up to a value of n − 1). . . .

"There are two more quantum numbers associated with each electron; the **magnetic quantum number** written as **m**, and the **spin quantum number**, written as **s**. . . .

"To make it easy to picture what is going on, the magnetic quantum number can be thought of as defining the amount of "tilt" there is to the orbit.

"The possible values for **m** follow the same rules as for **L**, except that negative numbers are now allowed (the "tilt" of the orbit can be either "up" or "down"). So for **n = 2**, the possible values for **m** would be 0, 1, or -1. . . .

"There are only two possible values for **s** [spin] for any value of **n**. These values are usually written as +1/2 and -1/2, meaning either a clockwise spin or an anticlockwise spin.

"But what do these numbers tell us about the electrons?

"Austrian physicist Wolfgang Pauli worked out the significance of these numbers in 1925. He suggested that no two electrons in any given atom could have exactly the same values for all four quantum numbers.

"This became known as the **Pauli exclusion principle – 'No two electrons in any atom may have the same set of quantum numbers'.**"[7]

E. Accuracies of the Models

The Neils Bohr Model was a close but not an exact fit to the measured data. The Sommerfeld Model of electron orbital ellipses, taking into account relativistic effects, gave a slightly better fit to the measured data. But as Thayer Watkins[8] demonstrates, no atomic structure model has a perfect fit to the measured data, and the Bohr Model is not much worse than the more advanced models. The fit is best between orbits of low quantum mechanical parameters n, and is worst between orbits of high quantum mechanical parameters n. At its best, the error can be as low as -0.01234 of 1%. But at its worst, the error can be at least as bad as -1.44546 of 1%.

These also are about the errors of the masses of charged leptons calculated in the next sections. Science has had 101 years to get the energies of the atom perfect, and has not done it. We should not hold back, therefore, until our model of energy states of semion orbits is perfect. We should publish a first cut mass model that is as close to the measured values as Bohr was to his measured values. The model we will advance in the next sections of the energies of semion orbits will be analogous to the Bohr model—not taking into account elliptical orbits, tilted orbits, or varying relativistic effects. There will be room for others to refine the model.

F. Differences of Semion Orbits with Electron Orbits

There are a number of differences between semion orbits and electron orbits:

1. Semions have e/2 charge. Electrons have whole e charge.
2. Each particle is a miniature black hole. The electron orbits as in Bohr's Model are exterior to black holes. The half charged particles called semions orbiting in charged leptons orbit inside black holes. This makes a difference in the force equation. The force for electrons in their orbit is $e^2/1 \cdot 4\pi\varepsilon_0 a^0 r^2$. The force on semions instead is $e^2/4 \cdot 4\pi\varepsilon_0 a^1 r^2$.
3. Semion orbits are a two body problem instead of a one body problem of electron orbits.
4. Semions orbit equal to or faster than the speed of light. Electrons orbit atoms slower than the speed of light.
5. The electron mass in the Bohr Model is the constant m_e. The semion mass in the outer non-relativistic frame is a variable.
6. Angular momentum in Bohr's atom is a function of n. But in our particles, the angular momentum is a function of not only n, but of $1/b$.
7. Between the models, standing waves have coincidence under different conditions. Instead of $C = n\lambda$ for electron orbits, $bC = n\lambda$ for semion orbits.
8. It is relatively easy to ionize electrons from orbit. It is virtually impossible to ionize semions from orbit. It is as though the semions are contained in a speed of light boundary which they cannot pass.
9. Because the semion orbits, with the addition of the gravitational aether velocity v, are faster than the speed of light c, the electric force between the semions is reversed in sense, and the sign on the potential energy is changed.

G. Deriving Particle States

Deriving particle states is one orbital level deeper than deriving electron orbital states that Niels Bohr did. The calculations are similar, but significantly different. Instead of treating the situation as a one body problem, we must treat the situation as a two body problem, with two equal semions in orbit about each other. This introduces an extra ½ into the expression of the centrifugal force.

We will solve for the particle states in a different order than Neils Bohr did. First we will balance the strong electric force in this problem with the central force of inertia in this problem, and familiarize ourselves with the whole problem to be solved:

Equation (3-16) below is the balancing of the force due to charge on the semions with the centrifugal force on the semions. The effective mass of a semion is half the mass of the whole particle in the outer non-relativistic frame. We use this mass of the semion in the centrifugal force along with the ½ from the two body problem. The velocity v_0 is greater than or equal to c, and must increase when the energy increases. In the electric type force side of the equation, the charge of the semion is e/2.

Different particle systems are in different order black holes. The force must depend on the order of black hole the particle system is in. Like the strong force and the electric force differ in strength by a power of $1/\alpha$, the forces in different orders of black holes differ by powers of $1/\alpha^{n/b}$. The electric type force expression, in the right side of Eq. (3-15), we expect to depend on a power of $1/\alpha^{n/b}$. To be in harmony with measured results and Eq. (21-15), we want the power of $1/\alpha$ to be related to n/b. Also, we want the power for the electron to be such that the power of α is 1 when n = 0. We therefore take the power of α for electrons and higher charged leptons to be n/b + 1. Completing the balancing of forces equation, we have

$$\frac{1}{2}\frac{m_e}{2}\frac{v_o^2}{r} = \left(\frac{e}{2}\right)^2 \frac{1}{4\pi\varepsilon_0 \alpha^{(n/b)+1}r^2} \tag{3-15}$$

The first ½ in the equation is from the two body nature of the problem, converting it to a one body problem. The $m_e/2$ is for the semion in ground state.   The v_o^2/r is the centrifugal force acting on the semions from a circular orbit. The $e/2$ is for the charge on each semion. The $4\pi\varepsilon_0$ are constants necessary to solve this problem in MKSC units.   The $1/\alpha^{n/b}$ is the fine structure coupling constant between the orders of black holes in the problem. The r is the radius of the particle sub-particle orbit.

Thanks to the two body nature of the problem, all of the numeral constants except 4 in the above equation cancel out. $e^2/4\pi\varepsilon_0\alpha$ can be factored out as ħc. One r can cancel out of the two sides of the equation. The equation then looks like the following:

$$m_e v_o^2 = \frac{\hbar c}{\alpha^{n/b}r} \tag{3-16}$$

We can solve for v_o if we can find an independent equation for r. Neils Bohr utilized the spin relation for electrons for this purpose. That equation was

$$m_e vr = n\hbar. \tag{3-17}$$

Unfortunately that spin relation does not work for charged leptons. Through trial and error, the author has settled on the following spin relation for charged leptons, which is here taken as a postulate:

$$m_e cr = n\hbar/b^2. \tag{3-18}$$

Solving for r we obtain:

$$r = n\hbar/b^2 m_e c. \qquad (3\text{-}19)$$

Combining Eqn. (3-19) with Eqn. (3-16) we have:

$$m_e v_o^2 = \hbar c b^2 m_e c / \alpha^{n/b} n\hbar \qquad (3\text{-}20)$$
$$v_o^2 = (b^2/n\alpha^{n/b}) c^2 \qquad (3\text{-}21)$$

$$v_o = (b^2/n\alpha^{n/b})^{1/2} c. \qquad (3\text{-}22)$$

The constant n in the above Eqn. is confusing. The n in Niels Bohr solution went 0, 1, 2, 3, 4, 5 . . . , but the n in the charged lepton solution goes 0, 1, 3, 6, 10, 15, 21 . . . , where b in the charged lepton solution goes 0, 1, 2, 3, 4, 5 . . . .

H. Deriving Semion Orbit Energy Levels and Masses

We have solved for v_o in terms of our parameters n and b. We can now plug that formula into the relationship for particle energy to obtain the energy levels of semion orbits, and thus the particle masses. The kinetic, potential, and total energies of the semion system can be expressed as

$$Energy_{total} = Energy_{kinetic} + Energy_{potential}. \qquad (3\text{-}23)$$

Because, in some particles, like charges attract, the potential energy for charged leptons is positive instead of negative for orbiting electrons in Hydrogen.

$$Energy_{total} = 1/2\, m_e v_o^2 + m_e v_o^2 \qquad (3\text{-}24)$$

$$= 3/2\, m_e v_o^2 \qquad (3\text{-}25)$$

Substituting Eq. (3-22) into Eq. (3-25), we obtain

$$\text{Energy}_{\text{total}} = 3/2 \, m_e \, (b^2/na^{n/b}) \, c^2. \qquad (3\text{-}26)$$

The measurable mass term of the semion system is

$$M_T = 3/2 \, (b_j^2/n_j \alpha^{n}{}_j{}^{/b}{}_j) \, m_e. \qquad (3\text{-}27)$$

The expression above in Eq. (3-27), derived from first principles, applies for a term in a series of terms for any charged lepton. But the mass of a particle equals that term plus a series of all previous terms back to that for the electron, where n and b equal zero (see Eq. (3-28)).

$$m_j = \frac{3}{2}\left\{\left(\frac{b_j^2}{n_j\alpha^{n_j/b_j}}\right) + \left(\frac{b_{j-1}^2}{n_{j-1}\alpha^{n_{j-1}/bj-1}}\right) + ...1\right\} m_e \quad (3\text{-}28)$$

The right most term in Eq. (3-28) times the g/2 factor for the electron is defined as 1.0.

To calculate this in general, we must have a definition of n, b, and j:

j	0	1	2	3	4	5 . . . .
n	0	1	3	6	10	15 . . . .
b	0	1	2	3	4	5 . . . .

Table 3-1

The first three n and b are tested. Higher n and b are calculated. We expect both n and b to increase with j. We expect $n_j - (n_{j-1})$ to be b_j.

Finally, just as the mass is a series of terms, all other force terms are added by multiplying by half the g-factor for the given particle. For the muon, the net mass is the sum of the last two terms according to Eq. (3-28) times half the g factor for the muon, or 206.553 998 611 59 m_e times $|g_\mu/2|$, or 206.553 998 611 59 m_e $|-1.001\ 165\ 912\ 4|$, equals 206.794 822 47. . . m_e, which is 1.000128, times the measured amount, 206.768 2843(52) m_e. That is 0.0128 of 1% error, which is almost the same error as the most accurate comparison of the theoretical Bohr Model of orbital differences to the measured values for orbital electrons. With what data we have to work with, our model is quite accurate.

There are only two usable measured g/2 factors that are available which can be used in calculating masses—for the electron and the muon. Fortunately, the match is close enough between particle masses and particle g/2 factors that calculated g/2 factors can be tested by the measured masses of the particles. We will begin employing calculated g/2 factors in this and the next two chapters.

I.   Fathoming the Orbital Velocities

Let us name the electron e_0, the muon e_1, and the tauon e_2. Then, for those particles and higher particles, Eq. (3-22) solves for the orbital semion velocities v_o:

Particle	Semion Orbital Velocity v_o
e_0	1.0000 c
e_1	11.7062 c
e_2	46.2480 c
e_3	167.8300 c
e_4	593.0600 c
e_5	2070.9000 c

Table 3-2

Compared to Einstein's Special Theory of Relativity, these are very high velocities. But these are velocities in a black hole. Velocities in a black hole should have no limit. Gravity escapes the bounds of a black hole, and communicates the sense of the mass of the black hole.

J.   Theorizing the Radii of Semion Orbits

This model also theorizes the radii of the semion orbits in charged leptons:

Particle	Radius of Semion Orbit r
e_0	1.0 ℏ/mc
e_1	1.0 ℏ/mc
e_2	3/4 ℏ/mc
e_3	6/9 ℏ/mc
e_4	10/16 ℏ/mc
e_5	15/25 ℏ/mc,

where m is the mass of the given charged lepton—not just the mass of the electron.

Table 3-3

K. Predicted Masses of Charged Leptons

This model theorizes and predicts the masses of any charged leptons. Data are calculated from CODATA[9] 2010 α and particle masses from the same source. Predicted mass includes (mass term plus all previous predicted mass ratios to mass of the electron in MeV) times (particle g/2 factor).

Charged Lepton	Mass Term (times m_e)	Predicted Mass	Measured Mass[9]
e_0	1.000 000 000	1.000 000 000	1.000 000 000
e_1	205.553 998 611	206.794 822 479	206.768 2843(52)
e_2	3 208.351 934	3 420.101 469	3 477.15
e_3	42 252.446 35	45 933.827 38	
e_4	527 591.655 3	577 827.418 5	

Table 3-4

This model does not predict a limited number of charged leptons (which we now observe). It predicts an infinite number of charged leptons, the next two of which are e_3 and e_4 in the above table.

The calculations in this chapter are for charged leptons. Similar calculations could be made for the quartons in the pion family. Other particle sets could be calculated by taking into account the Chonomic structures of the given particles. All the fundamental whole particles up to state 5 are calculated in the next chapter, from which all other particles up to state 5 may be calculated. Accompanying the calculation of fundamental masses in Chapter 4, is the calculation of the associated fundamental g/2 factors in Chapter 5.

---

[1]Stephen Gasiorowicz, *Quantum Physics* (New York: John Wiley & Sons, 1974), p. 15.

[2]*Ibid.*

[3]*Ibid.*

[4]"Bohr model," *Wikipedia*, the free encyclopedia, http://en.wikipedia.org/wiki/Bohr_model.

[5]*Ibid.*

[6]*Ibid.*

[7]Professor John Blamire, "Atomic Structure—The mystery of . . .—. . .the quantum atom," Exploring Life @ BIOdotEDU, http://www.brooklyn.cuny.edu/bc/ahp/LAD/C3/C3_elecPos_02.html.

[8]Thayer Watkins, "The Relativistic Bohr Model of a Hydrogen-like Atom," applet-magic.com: Silicon Valley, Tornado Alley & BB Island USA, http://www.applet-magic.com/relabohr.htm.

[9]http://physics.nist.gov/cuu/Constants/.

Problem Set 3

1. How many quantum numbers are there in the Electrino Model of mass calculations of charged leptons?

2. Does the author's model account for elliptical or tilted orbits?

3. How accurate is the author's prediction of the mass of the muon?

4. Why is the author's prediction of the mass of the tauon so inaccurate?

5. What charge do semions have?

6. What is the main difference between the force of electrons in their orbit and the force of semions in their orbit?

7. How many body problem are semions in orbit?

8. Do semions orbit slower than the speed of light or faster than the speed of light?

9. Do charged lepton semion orbital velocities follow Einstein's Special Theory of Relativity?

10. What difference in the mass is there between Bohr's Model and the author's model?

11. What different rule for the coincidence of standing waves in the particle does the author's model have as compared to Bohr's Model?

12. Are semions easy to ionize?

13. What reverses the sense of the potential energy in semion orbits?

14. What evidence is there that the mass in Eq. (3-18) is the overall mass of the charged lepton, not just the mass of the electron?

15. What forces are being balanced in Eq. (3-19)?

16. What factor occurs in the mass calculations due to both the kinetic and potential energies being positive?

17. What mass term for the charged lepton is derived from first principles?

18. How is that mass used in the calculation of the total mass of the particle?

19. What is n as a function of j? What is b as a function of j?

20. The Electrino Fusion Model of Elementary Particles predicts there will be how many different kinds of charged leptons in all?

21. This is a free question: Did you imagine that v_o should be so much above the speed of light for particles above electrons?

22. This is a free question: Are the radii of semion orbits a function of α? How?

Chapter 4

PREDICTION OF THE MASSES OF EVERY PARTICLE, STEP 1, REVISED

Gordon L. Ziegler and Iris Irene Koch

P.O. Box 1162, Olympia, WA 98507-1162 USA
e-mail ben_ent100@msn.com

PREFACE

The masses of charged leptons, anti-charged leptons, the pion family, the anti-pion family, and the neutron family are here calculated to state 4 or 5, and the way is thereby paved for the calculation of the masses of every known particle. This feat is not possible in the Quark Model, the Standard Model, the String Theory, or the Many-Dimensional Theory. It is possible only with the 'Electrino Hypothesis' that fractional charges come in $\pm e$, $\pm e/2$, $\pm e/4$, and $\pm e/8$, not the Quark Hypothesis that fractional charges come in $\pm 2e/3$ and $\pm e/3$. This electrino hypothesis is the basis for the far-reaching theory 'Electrino Fusion Model of Elementary Particles'. All that is used in this paper is the Electrino Hypothesis and algebra. All calculations are for either two-body problems or single-body problems. All particle bonds are seen to be orbital bonds.

The derivation from first principles of the masses of three or more states of particles in this paper is a great test of the Electrino Fusion Model of Elementary Particles.

In the theory here set forth, most electrinos (quartons, semions, unitons, and their anti-particles) are all trapped at or faster than the speed of light and cannot go slower than the speed of light, so cannot be detected directly.   Therefore the basis of the theory is more mathematical than physical, derived from first principles. Yet the model makes many physical predictions—the masses of many known particles.

1. Introduction

The masses of particles cannot be calculated in the Quark Model, The Standard Model, the String Theory, or the Many-Dimensional Theory.   This is not that the physicists have not yet figured out how to do the calculations in these models.   Rather, it is because the calculations are impossible in these models.   But it is possible to do them in a new Theory of Particle Physics— The Electrino Fusion Model of Elementary Particles.   That feat is accomplished in the present chapter, Chapter 5, and *Predicting Masses, by the author,* without tensors, matrices, Hamiltonians, Schrödinger's Equation, Isospin and many other advanced mathematical tools and concepts. This paper will use only algebra and the Electrino Hypothesis, which says that fractional charged particles come in $\pm e$, $\pm e/2$, $\pm e/4$, and $\pm e/8$, not in the $\pm 2e/3$ and $\pm e/3$ of the Quark Hypothesis.

The next thing to consider is that every known particle (except photons) can be constructed with various states of electrons, positrons, various states of pions, anti-pions, various states of neutrons, anti-neutrons and various combinations of those particles.   Because in the new theory there is a postulate that smooth symmetrical particles cannot have detectable spin, the theory does not allow electrons to be spinning point charges.   In the new theory, electrons are composed of two half particles (semions) orbiting about each other at the speed of light. Pions are composed of four fourth charges (quartons)—two orbiting one way, the other two orbiting the same way, and the two pairs of quartons orbiting the opposite way. A neutron is composed of a whole e particle (uniton) orbited by an electron (which is composed of two half charges orbiting about each other). Now if we could predict the masses of various states of electrons, various states of pions, and various states of neutrons, and their anti-particles, and learn how to put them together in compound particles, and learn how to calculate the masses of multi-particle particles, we would learn how to predict the masses of every known particle—which is precisely what we intend to do in this series of books and papers.

We first derived the prediction of the masses of charged leptons in [1]. That paper discussed the historical relationship to the Bohr's atom, and gave a list of the differences in the calculations. Here we attempt to abbreviate those calculations to facilitate understanding. In this chapter we take a different approach than in the last chapter. In the last chapter we took into account only one or two exponential polynomials characteristic per particle to calculate the masses of the particles. In this chapter, we take into account every sub-particle and every binding orbit in the calculation of the masses of the particles. In this chapter, the sub-particles are all electrinos—semions and quartons. When considered apart from their orbits, they amount to at rest. And at rest they have zero mass. They have mass only in their orbits.

2. ELECTRON FAMILY CALCULATIONS

Deriving particle states is one orbital level deeper than deriving electron orbital states that Niels Bohr did. The calculations are similar, but significantly different. Instead of treating the situation as a one body problem, we must treat the situation as a two body problem, with two equal semions in orbit about each other. This introduces an extra ½ into the expression of the centrifugal force.

We will solve for the particle states in a different order than Neils Bohr did. First we will balance the strong electric force in this problem with the central force of inertia in this problem, and familiarize ourselves with the whole problem to be solved:

Equation (4-1) below is the balancing of the force due to charge on the semions with the centrifugal force on the semions. The effective mass of a semion is half the mass of the whole particle in the outer non-relativistic frame. We use this mass of the semion in the centrifugal force along with the ½ from the two body problem. The velocity v_o is greater than or equal to c, and must increase when the energy increases. In the electric type force side of the equation, the charge of the semion is e/2.

Different particle systems are in different order black holes. The force must depend on the order of black hole the particle system is in. Like the strong force and the electric force differ in strength by a power of $1/\alpha$, the forces in different orders of black holes differ by powers of $1/\alpha^{n/b}$. The electric type force expression, in the right side of Eq. (4-1), we expect to depend on a power of $1/\alpha^{n/b}$. To be in harmony with measured results and Eq. (4-1), we want the power of $1/\alpha$ to be related to n/b. Also, we want the power for the electron to be such that the power of α is 1 when n = 0. We therefore take the power of α for electrons and higher charged leptons to be n/b + 1. Completing the balancing of forces equation, we have

$$\frac{1}{2}\frac{m_e}{2}\frac{v_o^2}{r} = \left(\frac{e}{2}\right)^2 \frac{1}{4\pi\varepsilon_0 \alpha^{(n/b)+1} r^2} \qquad (4\text{-}1)$$

The first ½ in the equation is from the two body nature of the problem, converting it to a one body problem. The $m_e/2$

is for the semion in ground state.   The v_o^2/r is the centrifugal force acting on the semions from a circular orbit. The e/2 is for the charge on each semion. The $4\pi\varepsilon_0$ are constants necessary to solve this problem in MKSC units.   The $1/\alpha^{n/b}$ is the fine structure coupling constant between the orders of black holes in the problem. The r is the radius of the particle sub-particle orbit.

Thanks to the two body nature of the problem, all of the numeral constants except 4 in the above equation cancel out. $e^2/4\pi\varepsilon_0\alpha$ can be factored out as $\hbar c$. One r can cancel out of the two sides of the equation. The equation then looks like the following:

$$m_e v_o^2 = \frac{\hbar c}{\alpha^{n/b} r} \qquad (4\text{-}2)$$

We can solve for v_o if we can find an independent equation for r. Neils Bohr utilized the spin relation for electrons for this purpose. That equation was

$$m_e v r = n\hbar. \qquad (4\text{-}3)$$

Unfortunately that spin relation does not work for charged leptons. Through trial and error, the author has settled on the following spin relation for charged leptons, which is here taken as a postulate:

$$m_e c r = n\hbar/b^2. \qquad (4\text{-}4)$$

Solving for r we obtain:

$$r = n\hbar/b^2 m_e c. \qquad (4\text{-}5)$$

Combining Eqn. (4-5) with Eqn. (4-2) we have:

$$m_e v_0^2 = \hbar c b^2 m_e c / \alpha^{n/b} n\hbar \qquad (4\text{-}6)$$

$$v_0^2 = (b^2/n\alpha^{n/b}) c^2 \qquad (4\text{-}7)$$

$$v_0 = (b^2/n\alpha^{n/b})^{1/2} c. \qquad (4\text{-}8)$$

The constant n in the above Eqn. is confusing. The n in Niels Bohr solution went 0, 1, 2, 3, 4, 5 . . . , but the n in the charged lepton solution goes 0, 1, 3, 6, 10, 15, 21 . . . , where b in the charged lepton solution goes 0, 1, 2, 3, 4, 5 . . . . The n's are calculable from the b's. $n_j = n_{j-1} + b_j$.

3. DERIVING SEMION ORBIT ENERGY LEVELS AND MASSES

We have solved for v_0 in terms of our parameters n and b. We can now plug that formula into the relationship for particle energy to obtain the energy levels of semion orbits, and thus the particle masses. The kinetic, potential, and total energies of the semion system can be expressed as

$$Energy_{total} = Energy_{kinetic} + Energy_{potential}. \qquad (4\text{-}9)$$

Because, in some particles, like charges attract, the potential energy for charged leptons is positive instead of negative for orbiting electrons in Hydrogen.

$$Energy_{total} = 1/2\ m_e v_0^2 + m_e v_0^2 \qquad (4\text{-}10)$$

$$= 3/2\ m_e v_0^2 \qquad (4\text{-}11)$$

Substituting Eq. (4-7) into Eq. (4-11), we obtain

$$Energy_{total} = 3/2\ m_e\ (b^2/n\alpha^{n/b}) c^2. \qquad (4\text{-}12)$$

The measurable mass term of the semion system is

$$m_T = 3/2 \ (b_j^2/n_j a^{n}{}_j{}^{/b}{}_j) \ m_e. \qquad (4\text{-}13)$$

     The expression above in Eq. (4-13), derived from first principles, applies for a term in a series of terms for any charged lepton. But the mass of a particle equals that term plus a series of all previous terms back to that for the electron, where n and b equal zero (see Eq. (4-14)).

$$m_j = \frac{3}{2} \left\{ \left(\frac{b_j^2}{n_j \alpha^{n_j / b_j}} \right) + \left(\frac{b_{j-1}^2}{n_{j-1} \alpha^{n_{j-1}/bj-1}} \right) + ...1 \right\} m_e \qquad (4\text{-}14)$$

The right most term in Eq. (4-14) times the g/2 factor for the electron is defined as 1.0.

     To calculate this in general, we must have a definition of n, b, and j:

j		0	1	2	3	4	5....
n		0	1	3	6	10	15....
b		0	1	2	3	4	5....

Table 4-1

The first three n and b are tested. Higher n and b are calculated. We expect both n and b to increase with j. We expect $n_j - (n_{j-1})$ to be b_j.

     Finally, just as the mass is a series of terms, all other force terms are added by multiplying by half the g-factor for the given particle. For the muon, the net mass is the sum of the last two terms according to Eq. (4-14) times half the g factor for the muon, or 206.553 998 611 59 m_e times $|g_\mu/2|$, or 206.553 998 611 59 m_e $|-1.001\ 165\ 912\ 4|$, equals 206.794 822 47. . . m_e, which is 1.000128, times the measured amount, 206.768 2843(52) m_e. That is 0.0128 of 1% error, which is almost the same error as the most accurate comparison of the theoretical Bohr Model of orbital differences to the measured values for orbital electrons. With what data we have to work with, our model is quite accurate.

There are only two usable measured g/2 factors that are available which can be used in calculating masses—for the electron and the muon. Fortunately, the match is close enough between particle masses and particle g/2 factors that calculated g/2 factors can be tested by the measured masses of the particles. We will begin employing calculated g/2 factors in this and the next chapter.

4. FATHOMING THE ORBITAL VELOCITIES

Let us name the electron e_0, the muon e_1, and the tauon e_2. Then, for those particles and higher particles, Eq. (4-8) solves for the orbital semion velocities v_0:

Particle	Semion Orbital Velocity v_0
e_0	1.0000 c
e_1	11.7062 c
e_2	46.2480 c
e_3	167.8300 c
e_4	593.0600 c
e_5	2070.9000 c

Table 4-2

Compared to Einstein's Special Theory of Relativity, these are very high speeds!

5. THEORETICAL RADII OF SEMION ORBITS

This model also theorizes the radii of the semion orbits in charged leptons:

Particle | Radius of Semion Orbit r

e_0	1.0 ℏ/mc
e_1	1.0 ℏ/mc
e_2	3/4 ℏ/mc
e_3	6/9 ℏ/mc
e_4	10/16 ℏ/mc
e_5	15/25 ℏ/mc,

Table 4-3

Only the first three particles have been observed. Higher state particles may be so small that the particles plow through each other, fusing to unitons.

6. PREDICTED MASSES OF CHARGED LEPTONS

This model predicts the masses of any charged leptons. Five calculated values are given in the next Table. The first three (e_0 through e_2) have been measured already, and are known to exist. The remaining two are calculations for particles which probably do not long exist. In the Table, measured $g/2$ factors are from reference [2] adapted; predicted $g/2$-factors are from the next chapter. The measured g/2-factors have error terms. The calculated do not.

charged lepton	b	n	predicted $3b^2 / 2n\alpha^{n/b}$	predicted m/m_e less g/2-factor	meas. or calc. g/2-factor
e_0	0	0	included	included	included
e_1	1	1	205.553 998	206.553 998	-1.001 165 920 7(06)
e_2	2	3	3,208.351 934	3,415.905 932	-1.001 157 653 136
e_3	3	6	42,252.446 35	45,875.90628	-1.001 165 744 345
e_4	4	10	527,591.655 3	577,091.0215	-1.001 167 869 444

Table 4-4

charged lepton	b	n	predicted m/m_e	measured m/m_e [8]
e_0	0	0	1.000 000 000	1.000 000 000
e_1	1	1	206.794 822 479	206.768 2843(52)
e_2	2	3	3,420.101 469	3,477.15(31)
e_3	3	6	45,933.827 38	
e_4	4	10	577,827.418 5	

Table 4-5

Only the first three particles have been observed. Higher state semions may be so small that they plow through each other, fusing to unitons.

7. MASSES OF THE POSITRON FAMILY

The positron family g/2 factors are the charge conjugates of the electron family g/2 factors, except they have additional terms for the meso-electric force. Otherwise, the mass tables for the positron family are the same as the mass tables for the electron family. We will denote the positron as $-e_0$.

charged lepton	b	n	predicted $3b^2 / 2n\alpha^{n/b}$	predicted m/m_e less g/2-factor	calculated g/2-factor
$-e_0$	0	0	1.000 000 000	1.000 000 000	+1.001 159 652 169
$-e_1$	1	1	205.553 998	206.553 998	+0.978 240 603 27
$-e_2$	2	3	3,208.351 934	3,415.905 932	+0.863 605 799 0
$-e_3$	3	6	42,252.446 35	45,875.906 28	+0.588 510 166 228
$-e_4$	4	10	527,591.655 3	577,091.0215	+0.084 145 505 542

Table 4-6

charged	b	n	predicted m/m_e	measured
				m/m_e [8]
lepton				
$-e_0$	0	0	-1.001 159 652	none
$-e_1$	1	1	-202.059 507 6	observed
$-e_2$	2	3	-2,949.996 103	yet
$-e_3$	3	6	-26,998.437 23	
$-e_4$	4	10	-48 559.615 75	

Table 4-7

8. NEW CALCULATIONAL TOOL

We wish to employ a new tool, never employed by the authors before. We wish to use it first with the electron. With the electron, semions orbit at c in the non-relativistic frame, and also orbit at c in the relativistic frame. In the non-relativistic frame, electrons also have a very small radial aether velocity v, and a radial aether speed c in the relativistic frame. Until now, the authors did not know how to calculate v, or know what it was. But v^2 is related to c^2 by the inverse square of the outer radius r_e, as compared to the inner radius R_0. In other words,

$$v^2 = -c^2 R_0^2 / r_e^2 \qquad (4\text{-}15)$$

What about higher members of the electron family? The orbital velocity squared v_{oj}^2 of the innermost mass term of an electron family member in the inner relativistic frame is

$$v_{oj}^2 = \left(b^2 / n\alpha^{n/b} \right) c^2 \quad . \qquad (4\text{-}16)$$

The orbital velocity squared in the innermost mass term of an electron family member in the outer non-relativistic frame is apparently the same. In fact, the inward and outward aether velocity squared at the surface of the orbit in the relativistic frame for the innermost mass term is apparently the same. The v^2 from the non-relativistic frame is yet to be determined. Let us try the inverse square relationship discovered above for the electron.

$$v_{rj}^2 = \frac{b_j^2 c^2}{n_j \alpha^{n_j/b_j}} \frac{(-)R_0^2 n_j \alpha^{n_j/b_j}}{b_j^2} \frac{1}{r_j^2} = -\frac{c^2 R_0^2}{r_j^2}. \qquad (4\text{-}17)$$

The radial aether velocity in the relativistic frame is affected by the increased mass, as is also the radius in the relativistic frame, but the radial velocity of the aether in the non-relativistic frame calculates simply with c, R_0, and r_j in Eq. (4-17). The radius r_j and velocity v_{rj} are different in each shell of mass. Therefore only an effective total aether velocity is associated with an effective radius of the whole particle.

Eqs. (4-15, 4-17) give us a new v^2 to solve gravitational problems with. We equate it to the v^2 in the relativistic approximation of the escape velocity:

$$v^2 = GM / r = -c^2 R_0^2 / r^2 \tag{4-18}$$

$$Mr = -c^2 R_0^2 / G = -c^2(-\hbar G) / G c^3 = \hbar / c \tag{4-19}$$

This is another important identity in particle physics.

We have now calculated from first principles the masses of e_3 and e_4—as yet undetected particles—as well as the electron, muon, and tauon. We could do the same for more particles. We turn now to the second portion of the calculation from first principles of the masses of every particle—the calculation of the masses of the pion family.

9. PION FAMILY CALCULATIONS

The first thing to consider in calculating the masses of the members of the pion family is the pion family calculations are very different from the electron family calculations. The electron family member has only one orbit of half particles (semions). A pion family member has three orbits in it—one of two fourth charges (quartons) orbiting one way, one of two other quartons orbiting the same way, and one of the two pairs of quartons (similar to semions) orbiting the opposite way.

The next thing to do is to calculate the velocity and velocity squared of each of the three orbits and a composite velocity for the whole system. Now let us repeat the calculation of the electron family members, making the changes necessary for the pion. Following the least action type calculations similar to what we did for the spin relation for the electron family, we find that whereas the spin relation was $r = n\hbar / mv$ for electrons orbiting atoms and $r = n\hbar / b^2 mc$ for the electron family of particles, the pion family has the spin relation $r = n^2 \hbar / bmc$. Therefore we take $bC = n^2 \lambda$ for the pion family members. This reduces to

$$\lambda = bC / n^2 = 2\pi br / n^2. \tag{4-20}$$

The energy of the particle system solves for

$$E = h\nu = 2\pi \hbar \nu = 2\pi \hbar c / \lambda = n^2 \hbar c / br = mc^2 \quad . \tag{4-21}$$

By the last equation in the above chain of equations, we see

$$r = n^2 \hbar / bmc \quad . \tag{4-22}$$

To Eq. (4-22) we add the balancing of the force due to charge on the quartons with the centrifugal force on the quartons. The effective mass of a quarton is one fourth of the mass of the whole particle in the outer non-relativistic frame. We use this mass of the quarton in the centrifugal force along with the $1/2$ from the two-body problem. The velocity v_0 is greater than c, and must increase when the energy increases. In the electric type force side of the equation, the charge of the quarton is $e/4$.

Each particle is a miniature mass singularity, and communicates with the outside world through powers of α (the Fine Structure Constant). The electric force expression, in the right side of Eq. (4-23), we expect to depend on a power of $1/\alpha$. The numerator of the power of alpha must be what makes the mass increase in the particle—namely the shells of mass from the radius r_j to $r \to \infty$, which can be totaled by taking $(b+1)$ (pairing of shells) times $b/2$ (number of pairs of shells). The denominator in the power of α should be b (the power of attenuation through j orders of mass singularity). Also, we want the power for the electron to be such that the power of α is 1 when $n = (b^2 + b)/2 = 0$. We take the power of α for electrons and higher charged leptons to be $n/b+1$. Balancing the forces, we have

$$\frac{1}{2}\frac{m}{4}\frac{v_0^2}{r} = (e/4)^2 \Big/ 4\pi\varepsilon_0 \alpha^{(n/b)+1} r^2 \quad . \tag{4-23}$$

The $1/2$ in the equation is from the two-body nature of the problem, converting it to a one-body problem. The first $1/4$ in the equation is for the quarton non-relativistic effective mass.

Some of the numeral constants in the above equation cancel out. The $e^2/4\pi\varepsilon_0\alpha$ can be factored out as $\hbar c$. One r can cancel out of the two sides of the equation. The equation then looks like the following:

$$mv_0^2 = \hbar c/2\alpha^{n/b} r, \text{ where } v_0 > c \quad \text{or} \quad n, b > 0 \quad . \tag{4-24}$$

Combining Eq. (4-24) with Eq. (4-22), we can solve for v_0:

$$v_0^2 = (b/2n^2\alpha^{n/b})c^2 \quad , \quad b, n > 0 \quad , \tag{4-25}$$

$$v_0 = \sqrt{b/2n^2\alpha^{n/b}}\; c \quad , \quad b, n > 0 \quad . \tag{4-26}$$

Both inner quarton orbits in pion family members have Eqs. (4-25, 4-26) as the solutions of the velocity and velocity squared for those orbits. Now let us determine the orbital velocity and velocity squared of the overall orbits of quarton pairs in the pion family members. We will use Eq. (4-22) again. But Eq. (4-23) is modified to Eq. (4-27).

$$\frac{1}{2}\frac{m}{2}\frac{v_0^2}{r} = (e/2)^2 \big/ 4\pi\varepsilon_0 \alpha^{(n/b)+1} r^2 \tag{4-27}$$

Reducing Eq. (4-27) similar to Eq. (4-23) and combining with Eq. (4-22) yields

$$v_0^2 = \left(b \big/ n^2 \alpha^{n/b}\right) c^2 \quad , \quad b, n > 0 \tag{4-28}$$

$$v_0 = \sqrt{b / n^2 \alpha^{n/b}}\, c \quad , \quad b, n > 0 \tag{4-29}$$

This is the velocity and velocity squared of the overall orbits of the quarton pairs in the pion family members.

Now we need to arrive at a composite orbital velocity and velocity squared for the whole quarton system. Since the centers of the two quarton orbits orbit at right angles to the velocities of the quarton orbits at those points, we can treat the velocities at right angles to each other, and add the squares of the velocities to obtain the total v_T^2. The velocities squared of the inner quarton orbits are each half the magnitude of the velocity squared of the overall orbit. The sum of the square of the three velocities, then, is

$$v_T^2 = \left(2b / n^2 \alpha^{n/b}\right) c^2 \quad , \quad b, n > 0 \tag{4-30}$$

$$v_T = \sqrt{2}\sqrt{b / n^2 \alpha^{n/b}}\, c \quad , \quad b, n > 0 \tag{4-31}$$

10. DERIVING PION FAMILY ENERGY LEVELS AND MASSES

We have solved for v_T in terms of our parameters n and b which we can insert in energy Eq. (4-34). The mass m in Eq. (4-33) and subsequent calculated masses depend on v_T, but m_v (the mass multiplying v_0^2 in the calculations) does not increase with v_0 as in relativity.

This is a non-relativistic calculation, and $m_v = m_e$. [See Eq. (4-34).] The other parameters are not interchangeable, but come in matched sets of subscript j. Thus kinetic, potential, and total energies of the semion system can be expressed as

$$\text{Energy}_{\text{total}} = \text{Energy}_{\text{kinetic}} + \text{Energy}_{\text{potential}} \qquad (4\text{-}32)$$

Though we have already calculated v_T^2, and can insert that in Eq. (4-34), it is interesting to see equivalent calculations in terms of the potential energy of the overall orbit of quarton pairs:

$$\frac{1}{4}E_p + \frac{1}{4}E_p - \frac{1}{2}E_p + \frac{1}{2}E_p + \frac{1}{2}E_p + \frac{1}{1}E_p = 2E_p = mc^2 , (4\text{-}33)$$

where the first three terms are kinetic energies, and the last three terms are potential energies of the orbits, where Eq. (4-22) is substituted for r in Eq. (4-24) to obtain the potential energy fraction in Eq. (4-33), and where

$$E_p = m_v v_0^2 = (b / n^2 \alpha^{n/b}) m_v c^2 = (b / n^2 \alpha^{n/b}) m_e c^2 \qquad (4\text{-}34)$$

The calculable mass term of the quarton pair orbiting quarton pair system (less the g/2 factor) is

$$m_j = \left(2b_j / n_j^2 \alpha^{n_j b_j}\right) m_e \quad . \qquad (4\text{-}35)$$

Mass is a volume thing, and is integrated from $r = r_j$ to $r \rightarrow \infty$ in discrete terms. The expression above in Eq. (4-35), derived from first principles, is a term in a series of terms in a natural calculation of the mass of the pion family member. The sum of the terms is equal to m.

$$m = m_j + m_{j-1} + m_{j-2} + \ldots + m_\pi \quad . \qquad (4\text{-}36)$$

For the pion, $m = m_\pi$. For the kaon, there is one term besides the pion, where $b_j = 2$. For the D-on, there are two terms beside the pion, *etc* (this is without the g/2-factors).

The first author thought there should be no g/2-factors for the pion family members. But there must be $g/2$-factors after all for the pion family members. The strong term must be +1.0 instead of -1.0. And the most variable term is a term for the meso-electric force, not included in the electron $g/2$-factor. It should be

$$-(p-1)n_{p-1}\pi\alpha.$$

We can now write an equation for the mass of the term j of the pion family members, taking into account the contribution of each quarton orbit and the overall orbit of the quarton pairs, but not counting the $g/2$-factors. We take care to differentiate m_j from m. We now have

$$m_j = 2\,b_j m_{\rm e}\big/ n_j^2 \alpha^{n_j/b_j} \qquad . \qquad (4\text{-}37)$$

11. PREDICTIONS OF THE MASSES OF THE PION FAMILY

We now present Table 4-8 of the predicted (calculated from first principles) and the measured values of the masses of the pion family. [5] In the Table, pions are denoted by π_1, kaons by π_2, D-ons by π_3, etc.

Particle	b	n	Calculated $2b/n^2\alpha^{n/b}$	Predicted m/m_e less $g/2$-factor	Calculated [7] $g/2$-factor
Pion π_1	1	1	274.071 998	274.071 998	+1.001 157 533
Kaon π_2	2	3	712.967 096 5	987.039 094	+0.978 240 603
D-on π_3	3	6	3,129.810 84	4,390.921932	+0.863 605 778
π_4	4	10	17,586.388 5	22,251.382 4	+0.588 510 179

Particle	b	n	Predicted m / m_e	Measured m / m_e [8]
Pion π_1	1	1	274.389 245	273.132 04
Kaon π_2	2	3	965.561 718	966.101 8
D-on π_3	3	6	3,792.025 5	3,658.755
π_4	4	10	13,095.165	

Table 4-9

12. MASSES OF THE ANTI-PION FAMILY

The anti-pion family g/2 factors are the charge conjugates of the pion family g/2 factors, except they do not have terms for the meso-electric force. Otherwise, the mass tables for the anti-pion family are the same as the mass tables for the pion family. We will denote the anti-pion as $-\pi_1$.

Particle	b	n	Calculated $2b / n^2 \alpha^{n/b}$	Predicted m / m_e less $g / 2$-factor	Calculated [7] $g / 2$-factor
Anti-Pion - π_1	1	1	274.071 998	274.071 998	-1.001 157 533
Anti-Kaon - π_2	2	3	712.967 096	987.039 094	-1.001 165 912
Anti-D-on - π_3	3	6	3,129.810 84	4,390.111 09	-1.001 157 653
- π_4	4	10	17,586.388 5	23,237.6106	-1.001 165 744

Table 4-10

Particle		b	n	Predicted m/m_e	Measured* m/m_e [8]
Anti-Pion	$-\pi_1$	1	1	274.389 245	not
Anti-Kaon	$-\pi_2$	2	3	988.189 894	observed
Anti-D-on	$-\pi_3$	3	6	4,395.193 1	yet
	$-\pi_4$	4	10	23,264.699	

Table 4-11

13. NEUTRON FAMILY CALCULATIONS

The third particle type—the neutron family—is different from the other two particle types. The neutron is a baryon, and, like all baryons, has affecting the mass calculations not only a relativistic imaginary-axis massive core, but also at the same time a real-axis zero mass core—the uniton. Electron family members orbit about this massive core particle. It is similar to electrons orbiting protons in Hydrogen. But electrons orbiting a proton are easily ionized, whereas the electron orbiting the uniton in a neutron is strongly bound to the core uniton. The uniton cannot come alone. For all lower particles, an electron always accompanies the uniton. Electrons orbiting protons could have semions in higher states, but this is not normally considered. The semions in the electron orbiting the uniton in neutrons, however, provide neutrons with possible higher states for the neutron. Therefore we consider them in the neutron family calculations.

Just as each previous particle family type has a different spin relation, the electron orbits of the uniton have a different spin relation. With the neutron family orbits, we will use B and N for the whole particle electron orbiting the uniton, to differentiate it from b and n in the electron family orbits (also used in these calculations). The total neutron spin is mv_0r, but the observable spin on the event

horizon of the black hole is only mcr. We have to go by
the observable spin. The neutron observable spin relation
is:

$$mcr = N\hbar/B \tag{4-38}$$

$$r = N\hbar/Bmc \tag{4-39}$$

This type of spin relation also overrules the electron family
spin relation with b and n in the neutron.

Let us balance the force equation for the neutron
family. Instead of the balancing of the electric force and
the inertial force in Eq. (4-40) starting with a 1/2, as in Eqs.
(4-1), (4-23), and (4-27), because of the two-body nature of
those problems, Eq. (4-40) starts with a '1', because of the
single body nature of the neutron family problem. In this
problem, the mass is m/1 instead of m/2, because in the
main electron family orbits about the uniton, we are dealing
with electrons as whole particles, not half particles.

$$1(m/1)v_0^2/r = (e/1)^2/4\pi\varepsilon_0\alpha^{(N/B)+1}r^2 \tag{4-40}$$

Using the techniques under Eq. (4-1), this reduces to

$$mv_0^2 = \hbar c/\alpha^{N/B}r \quad . \tag{4-41}$$

Combining this with Eq. (4-39), we obtain

$$mv_0^2 = \hbar cBmc/N\hbar\alpha^{N/B} \tag{4-42}$$

$$v_{on}^2 = B/N\alpha^{N/B} c^2 \tag{4-43}$$

This is the orbital velocity squared for the neutron orbit of
the electron family member. We have to combine that with
the orbital velocity squared for the electron family member
v_{oe}^2, which is solved by substituting b for B and n for N in
Eqn. (4-43), because the neutron space charge limits the

spin relation for the intrinsic electron family members in the neutron, and is

$$v_{oe}^2 = b/na^{n/b} \, c^2 \qquad (4\text{-}44)$$

There is a region where v_{on} relative to v_{oe} is faster, and a region where it is slower, but the average of the absolute value v_{on} and average absolute value v_{oe} are at right angles to each other and can be added by squaring them. The process is clarified in Eqn. (4-45).

$$v_{Tn}^2 = [(B/Na^{N/B}) + (b/na^{n/b})] \, c^2 \qquad (4\text{-}45)$$

By multiplying the quantity in Eq. (4-45) by m_e we obtain the potential energy of the neutron family system. By multiplying the potential energy of the neutron family members by $3/2$, we obtain the total energy including the kinetic energy in the neutron. By dividing by c^2 we obtain the mass m of the neutron family member. Each particle has a $g/2$-factor. [7] The result is in Eq. (4-46).

$$m_{particle} = 3/2 \, [(B_i/Na^{N/B} \, g_i/2) + (b_j/na^{n/b} \, g_j/2)] \, m_e \qquad (4\text{-}46)$$

Unitons are different from other whole-body systems. There are no elevated states of unitons. They are always only at state 2. The only elevated states associated with unitons are with orbiting particle systems surrounding the unitons. This feature of unitons apparently is reflected in the property that uniton systems have only one shell of mass. The sum of the velocity and thus mass components of the electron family member and the neutron system is not compounded by layers of mass shells. That typical stage of calculations will be left out of baryon calculations.

We see the mass of the neutron is a combination of two calculable terms. The N's and the n's are calculable from the B's and the b's [see the paragraph after Eqn. (4-8)]. For all neutron family members, the orbital spin is -1. The B for all neutron family members is 2. On the other hand, the minimum b and n are 0. The neutron family members all have J = ½ and parity +. They all have unitons for core particles. They all have an electron family member with $\hbar/2$ intrinsic spin orbiting around the uniton with $\hbar$ orbital spin. The only things that differentiate the neutron family members are the energy states of the particles. Yet all the observed particles with these properties have different, seemingly unrelated, traditional names. Those observed so far are n, Σ^0, and Λ_b^0. To show the neutron related nature of those particles, we shall name those same particles $n = n_1$, $\Lambda = n_2$, $\Sigma^0 = n_3$, and $\Lambda_b^0 = n_4$, , etc.

The g/2 terms are simplified also. There is only one g/2 factor for the B terms—at B = 2 (Λ g/2 factor, see Chapter 5). For the intrinsic states of the semion orbits in the orbiting electron about the uniton, see the electron family g/2 factors in Chapter 5.

With the neutron family members, there is too much information to put in one table. We will divide the information into three tables.

Particle		B	N	$B/N\alpha^{N/B}$	b	n	$b/n\alpha^{n/b}$
n	n_1	2	3	1,069.450 645	0	0	1.000 000 000
Λ	n_2	2	3	1,069.450 645	1	1	137.035 999 7
Σ^0	n_3	2	3	1,069.450 645	2	3	1,069.450 645
Λ_b^0	n_4	2	3	1,069.450 645	3	6	9,389.432 523

Table 4-12 Neutron family parameters

In the next table, the g/2 factor to be utilized for B = 2 and N = 3 is the one where the meso-electric factor is -2 x 3πα, or the g/2 factor for the Λ particle.

Parti-cle	B_i	N_i	$g_i/2$	b_j	n_j	$g_j/2$
n	2	3	-1.138 716 794	0	0	-1.000 000 000
Λ	2	3	-1.138 716 794	1	1	-1.001 165 912
Σ^0	2	3	-1.138 716 794	2	3	-1.001 157 653
Λ_b^0	2	3	-1.138 716 794	3	6	-1.001 165 744

Table 4-13 System g/2 factors

Particle	Predicted Mass Ratio (m/m_e)	Measured Mass Ratio [8] (m/m_e)
n	1,828.202 115	1,838.683 66
Λ	2,032.495 772	2,183.337
Σ^0	2,288.490 108	2,333.9
Λ_b^0	10,618.179 61	10,996

Table 4-14

The calculated mass ratios of each neutron family member are a little on the low side compared to the measured values. This is to be expected because we are calculating only circular and not elliptical orbits and neglecting relativistic effects. But actually, our calculated values are a pretty good fit with the measured values. When our calculated values go up by small steps, the measured values go up by small steps. And when our calculated value goes up by a large step, the measured value goes up by a large step—with about the same degree of precision.

14. MASSES OF THE ANTI-NEUTRON FAMILY

There are no observed masses of the anti-neutron family to compare with. Therefore we will not do this one.

15. FORWARD LOOK

The neutron is a multi-particle particle. And we have done pretty well with the neutron family. This gives us hope that this science of chonomics, with the Unified Field Theory, can do well also with the proton and all other multi-particle particles. But for those calculations, please see the upcoming book, *Predicting the Masses,* by Gordon L. Ziegler and Iris Irene Koch, now a work in process.

16. CONCLUSION

This chapter has predicted the masses of five charged leptons, five anti-charged leptons, four members of the pion family, four members of the anti-pion family, as well as four members of the neutron family—22 predicted particle masses to two or four place accuracy. With the exception with the neutron family, that may be more than actually exist, because if the particle masses get too high, they can fuse to higher fusion state particles, and no longer be in the same family of particles. This bears closer observations in future particle physics. The formulas and tables show how to predict the masses of more particles of the electron and pion families if they exist. No other model of physics can do this. Therefore the Electrino Model of Elementary Particles should be carefully considered, and the further tests, described in Ref. [9], Chapter. 9, be funded and carried out.

For $70 million or less, a plethora of well explained scientific "miracles" could be tested. Is that too good to be true? Or is it too good to be false? Don't we need to find out?

Miraculous Effects of the Refresher

Reverse aging for adults
The simplest effect of the Refresher to understand is reversing adult aging. Old people can be made young adults again in the active footprint of the Refresher. This effect for positron anti-semion fusion does not really back up time or the clock. It merely reverses the order to disorder arrow in adults. It saturates at the maximum state of order—which is young adulthood. It reverses adult aging at a rate of about 1836 times as fast as the rate the original adult aging occurred. A century of aging can be reversed in just under 20 days.

Resurrections from the dead
The reverse aging occurs also for bodily remains— re-assembling dust and bones into living beings again. All the dead of all ages of earth's history would be resurrected in about 3½ years of Refresher machine time, starting with those who died most recently.

Backing diseases out of existence
In the process of reverse aging, diseases would be backed out of existence. This would work also for difficult diseases like HIV AIDS, cancer, and cystic fibrosis.

Reversing all decay
Spoiled fruit would un-spoil in the active footprint of the Refresher. Fresh fruit would stay at the maximum state of order for fruit forever—fresh picked fruit. And this would be without refrigeration. This would amount to a new kind of food preservation without canning or freezing.

This process would un-decay everything in the Refresher footprint, not just fruit. And the footprint could be enlarged to cover the entire earth.

Reversing pollution out of existence

In the Refresher footprint, all pollution would be reversed out of existence. Depending on the Refresher control settings, this effect could be world-wide.

"Raising up the foundations of many generations" Isaiah 58:12.

The Refresher would automatically rebuild previous decayed structures. It would rebuild and restore the entire earth.

Reversing forest fires

The Refresher not only would stop forest fires in its footprint, but would reverse the fires—restoring all that was lost—animate and inanimate, including lost trees and homes.

Reversing all calamities;
Reversing all effects of war;
Preventing all munitions from firing;
"Making wars to cease to the end of the earth." Psalm 46:9.
Removing sinful propensities from people, including criminals;
Emptying prison houses;
Making possible and efficient Clean Energy Sources.
The blessings of the Refresher are endless. In short it would restore earth to Edenic perfection in about 3½ years of machine time.

REFERENCES

[1] Gordon L. Ziegler and Iris Irene Koch, "Prediction of the Masses of Charged Leptons," Galilean Electrodynamics, November/December 2009, Vol. 20, No. 6, pp. 114-118.

[2] The NIST Reference on Constants, Units, and Uncertainty, CODATA Internationally recommended values of the Fundamental Physical Constants, Latest (2010) values of the constants, http://physics.nist.gov/cuu/Constants/.

[3] Thayer Watkins, "The Relativistic Bohr Model of a Hydrogen-like Atom," applet-magic.com: Silicon Valley, Tornado Alley & BB Island USA, http://www.applet-magic.com/relabohr.htm.

[4] **CRC Handbook of Chemistry and Physics**, 85th Edition, 2004-2005 (CRC Press, Boca Raton, 2004), p. **11-3**.

[5] **CRC Handbook of Chemistry and Physics**, 80th Edition, 1999-2000 (CRC Press, Boca Raton, 1999), pp. **11-1** to **11-45**.

[6] G.L. Ziegler, "A New Way to Calculate Electron and Muon g/2-factors," Galilean Electrodynamics, January/February 2006, Vol. 17, No. 1, pp. 11-15.

[7] This book, Chapter 5.

[8] J. Beringer *et al.* (Particle Data Group), PR **D86**, 010001 (2012) and 2013 partial update for the 2014 edition (URL: http://pdg.lbl.gov).

[9] Gordon L. Ziegler and Iris Irene Koch, *Electrino Physics Draft 2*, Chapter 9.

Chapter 5

PATTERN g/2 FACTORS

This chapter was calculated before the advanced light in Chapter 4 on the method of the calculation of the neutron family particle mass ratios was determined. As it turned out, we only needed one g/2 factor for the electrons orbiting the uniton. But instead of gutting the g/2 factors in in this chapter for the neutron and anti-neutron families, to match the Chapter 4 results, we preserve all the original g/2 factor calculations in this chapter in case we need them in calculating subsequent multi-particle particles, for the g/2 factors are very difficult to calculate, and we do not want to have to calculate them again.

This chapter succeeds "An Update on g/2 Factors" in the original Chapter 5 in *Advanced Electrino Physics*. That paper was a timely advance in the field, but it contained errors. Also it was incomplete. It had g/2 factors for fundamental matter particles—the electron family, the pion family, and the neutron family. But it did not include g/2 factors for their anti-particles, which, because of the meso-electric terms for positive but not negative particles, are not simple charge conjugates to the matter particles. This chapter corrects the original matter g/2 factors in Chapter 5, and adds the corresponding g/2 factors for the antimatter fundamental particles. With all the g/2 factors in this chapter, you can calculate any particle mass up to state 4 or 5 in the electron, pion, and neutron families and their anti-particles.

Table 5-1
Electron family
Electron e_0 g/2 Factor Evaluation with 2010 α

force:	g/2 factor term:	numerical value:
strong	-1	$-1.000\ 000\ 000\ 000$
electric	$-\alpha/2\pi$	$-0.001\ 161\ 409\ 733$
magnetic	$-\alpha^2/16\pi^2$	$-0.000\ 000\ 337\ 218$
$weak_1$	$+\alpha^2/8\pi$	$+0.000\ 002\ 118\ 804$
$weak_2$	$-\alpha^3/4\pi$	$-0.000\ 000\ 030\ 923$
$weak_3$	$+(32\alpha)^1\,\alpha^3/4\pi$	$+0.000\ 000\ 007\ 221$
$weak_4$	$-(32\alpha)^3\,\alpha^3/4\pi$	$-0.000\ 000\ 000\ 393$
$weak_5$	$+(32\alpha)^4\,\alpha^3/4\pi$	$+0.000\ 000\ 000\ 091$
$weak_6$	$-(32\alpha)^5\,\alpha^3/4\pi$	$-0.000\ 000\ 000\ 021$
$weak_7$	$+(32\alpha)^6\,\alpha^3/4\pi$	$+0.000\ 000\ 000\ 005$
$weak_8$	$-(32\alpha)^7\,\alpha^3/4\pi$	$-0.000\ 000\ 000\ 001$
$weak_9$	$+(32\alpha)^8\,\alpha^3/4\pi$	$+0.000\ 000\ 000\ 000$
total calculated $g_e/2$		$-1.001\ 159\ 652\ 169$
compare measured $g_e/2$		$-1.001\ 159\ 652\ 181\ 1(08)$ [1]

Table 5-2
Electron family
Muon e_1 g/2 Factor Evaluation with 2010 α

force:	g/2 factor term:	numerical value:
strong	-1	-1.000 000 000 000
electric	$-\alpha / 2\pi$	-0.001 161 409 733
magnetic	$-\alpha^2 / 16\pi^2$	-0.000 000 337 218
$weak_1$	$-\alpha^2 / 4\pi$	-0.000 004 237 608
$weak_2$	$+3\alpha^3 / 4\pi$	+0.000 000 092 769
$weak_3$	$-3(32\alpha)^1 \alpha^3 / 4\pi$	-0.000 000 021 663
$weak_4$	$+3(32\alpha)^3 \alpha^3 / 4\pi$	+0.000 000 001 181
$weak_5$	$-3(32\alpha)^4 \alpha^3 / 4\pi$	-0.000 000 000 275
$weak_6$	$+3(32\alpha)^5 \alpha^3 / 4\pi$	+0.000 000 000 064
$weak_7$	$-3(32\alpha)^6 \alpha^3 / 4\pi$	-0.000 000 000 015
$weak_8$	$+3(32\alpha)^7 \alpha^3 / 4\pi$	+0.000 000 000 003
$weak_9$	$-3(32\alpha)^8 \alpha^3 / 4\pi$	-0.000 000 000 000
$weak_{10}$	$+3(32\alpha)^9 \alpha^3 / 4\pi$	+0.000 000 000 000

total calculated g/2 factor for muon   -1.001 165 912 495
compare measured g/2 factor muon   -1.001 165 920 7(06)

Table 5-3
Electron family
Tauon e_2 g/2 Factor Evaluation with 2010 α

force:	g/2 factor term:	numerical value:
strong	-1	$-1.000\ 000\ 000\ 000$
electric	$-\alpha/2\pi$	$-0.001\ 161\ 409\ 733$
magnetic	$-\alpha^2/16\pi^2$	$-0.000\ 000\ 337\ 218$
$weak_1$	$+\alpha^2/4\pi$	$+0.000\ 004\ 237\ 608$
$weak_2$	$-6\alpha^3/4\pi$	$-0.000\ 000\ 185\ 539$
$weak_3$	$+6(32\alpha)^1\,\alpha^3/4\pi$	$+0.000\ 000\ 043\ 663$
$weak_4$	$-6(32\alpha)^3\,\alpha^3/4\pi$	$-0.000\ 000\ 002\ 362$
$weak_5$	$+6(32\alpha)^4\,\alpha^3/4\pi$	$+0.000\ 000\ 000\ 551$
$weak_6$	$-6(32\alpha)^5\,\alpha^3/4\pi$	$-0.000\ 000\ 000\ 128$
$weak_7$	$+6(32\alpha)^6\,\alpha^3/4\pi$	$+0.000\ 000\ 000\ 030$
$weak_8$	$-6(32\alpha)^7\,\alpha^3/4\pi$	$-0.000\ 000\ 000\ 007$
$weak_9$	$+6(32\alpha)^8\,\alpha^3/4\pi$	$+0.000\ 000\ 000\ 001$
$weak_{10}$	$-6(32\alpha)^7\,\alpha^3/4\pi$	$-0.000\ 000\ 000\ 000$

total calculated g/2 factor for tauon $-1.001\ 157\ 653\ 136$

Table 5-4
Electron family
e_3 g/2 Factor Evaluation with 2010 α

force:	g/2 factor term:	numerical value:
strong	-1	-1.000 000 000 000
electric	$-\alpha / 2\pi$	-0.001 161 409 733
magnetic	$-\alpha^2 / 16\pi^2$	-0.000 000 337 218
$weak_1$	$-\alpha^2 / 4\pi$	-0.000 004 237 608
$weak_2$	$+10\alpha^3 / 4\pi$	+0.000 000 309 233
$weak_3$	$-10(32\alpha)^1 \alpha^3 / 4\pi$	-0.000 000 072 210
$weak_4$	$+10(32\alpha)^3 \alpha^3 / 4\pi$	+0.000 000 005 937
$weak_5$	$-10(32\alpha)^4 \alpha^3 / 4\pi$	-0.000 000 001 919
$weak_6$	$+10(32\alpha)^5 \alpha^3 / 4\pi$	+0.000 000 000 214
$weak_7$	$-10(32\alpha)^6 \alpha^3 / 4\pi$	-0.000 000 000 050
$weak_8$	$+10(32\alpha)^7 \alpha^3 / 4\pi$	+0.000 000 000 011
$weak_9$	$-10(32\alpha)^8 \alpha^3 / 4\pi$	-0.000 000 000 002
$weak_{10}$	$+10(32\alpha)^9 \alpha^3 / 4\pi$	+0.000 000 000 000
$weak_{11}$	$-10(32\alpha)^{10} \alpha^3 / 4\pi$	-0.000 000 000 000

total calculated g/2 factor for e_3 -1.001 165 744 345

Table 5-5
Electron family
e$_4$ g/2 Factor Evaluation with 2010 α

force:	g/2 factor term:	numerical value:
strong	-1	-1.000 000 000 000
electric	$-\alpha/2\pi$	-0.001 161 409 733
magnetic	$-\alpha^2/16\pi^2$	-0.000 000 337 218
$weak_1$	$+\alpha^2/4\pi$	+ 0.000 004 237 608
$weak_2$	$-15\alpha^3/4\pi$	-0.000 000 463 849
$weak_3$	$+15(32\alpha)^1\alpha^3/4\pi$	+0.000 000 103 316
$weak_4$	$-15(32\alpha)^3\alpha^3/4\pi$	-0.000 000 005 633
$weak_5$	$+15(32\alpha)^4\alpha^3/4\pi$	+0.000 000 001 315
$weak_6$	$-15(32\alpha)^5\alpha^3/4\pi$	-0.000 000 000 307
$weak_7$	$+15(32\alpha)^6\alpha^3/4\pi$	+0.000 000 000 071
$weak_8$	$-15(32\alpha)^7\alpha^3/4\pi$	-0.000 000 000 016
$weak_9$	$+15(32\alpha)^8\alpha^3/4\pi$	+0.000 000 000 003
$weak_{10}$	$-15(32\alpha)^9\alpha^3/4\pi$	-0.000 000 000 000
$weak_{11}$	$+15(32\alpha)^{10}\alpha^3/4\pi$	+0.000 000 000 000

total calculated g/2 factor for e$_4$ -1.001 167 869 444

Table 5-6
Positron family
Anti-electron -e_0 g/2 Factor Evaluation with 2010 α

force:	g/2 factor term:	numerical value:
strong	$+1$	$+1.000\ 000\ 000\ 000$
meso-electric	$-bn\pi\alpha$	$-0.000\ 000\ 000\ 000$
electric	$+\alpha/2\pi$	$+0.001\ 161\ 409\ 733$
magnetic	$+\alpha^2/16\pi^2$	$+0.000\ 000\ 337\ 218$
$weak_1$	$-\alpha^2/8\pi$	$-0.000\ 002\ 118\ 804$
$weak_2$	$+\alpha^3/4\pi$	$+0.000\ 000\ 030\ 923$
$weak_3$	$-(32\alpha)^1\,\alpha^3/4\pi$	$-0.000\ 000\ 007\ 221$
$weak_4$	$+(32\alpha)^3\,\alpha^3/4\pi$	$+0.000\ 000\ 000\ 393$
$weak_5$	$-(32\alpha)^4\,\alpha^3/4\pi$	$-0.000\ 000\ 000\ 091$
$weak_6$	$+(32\alpha)^5\,\alpha^3/4\pi$	$+0.000\ 000\ 000\ 021$
$weak_7$	$-(32\alpha)^6\,\alpha^3/4\pi$	$-0.000\ 000\ 000\ 005$
$weak_8$	$+(32\alpha)^7\,\alpha^3/4\pi$	$+0.000\ 000\ 000\ 001$
$weak_9$	$-(32\alpha)^8\,\alpha^3/4\pi$	$-0.000\ 000\ 000\ 000$

total calculated g/2 factor for $-e_0$ $+1.001\ 159\ 652\ 169$

Table 5-7
Positon family
Anti-muon -e_1 g/2 Factor Evaluation with 2010 α

force:	g/2 factor term:	numerical value:
strong	$+1$	+1.000 000 000 00
meso-electric	$-bn\pi\alpha$	-0.022 925 309 22
electric	$+\alpha / 2\pi$	+0.001 161 409 73
magnetic	$+\alpha^2 / 16\pi^2$	+0.000 000 337 21
$weak_1$	$+\alpha^2 / 4\pi$	+0.000 004 237 60
$weak_2$	$-3\alpha^3 / 4\pi$	-0.000 000 092 76
$weak_3$	$+3(32\alpha)^1 \alpha^3 / 4\pi$	+0.000 000 021 66
$weak_4$	$-3(32\alpha)^3 \alpha^3 / 4\pi$	-0.000 000 001 18
$weak_5$	$+3(32\alpha)^4 \alpha^3 / 4\pi$	+0.000 000 000 27
$weak_6$	$-3(32\alpha)^5 \alpha^3 / 4\pi$	-0.000 000 000 06
$weak_7$	$+3(32\alpha)^6 \alpha^3 / 4\pi$	+0.000 000 000 01
$weak_8$	$-3(32\alpha)^7 \alpha^3 / 4\pi$	-0.000 000 000 00
$weak_9$	$+3(32\alpha)^8 \alpha^3 / 4\pi$	+0.000 000 000 00
$weak_{10}$	$-3(32\alpha)^9 \alpha^3 / 4\pi$	-0.000 000 000 00

total calculated g/2 factor for $-e_1$ +0.978 240 603 27

* b is advanced 1.

Table 5-8
Positron family
Anti-tauon $-e_2$ g/2 Factor Evaluation with 2010 α

force:	g/2 factor term:	numerical value:
strong	$+1$	+1.000 000 000 0
meso-electric	$-bn\pi\alpha$	-0.137 551 855 3
electric	$+\alpha/2\pi$	+0.001 161 409 7
magnetic	$+\alpha^2/16\pi^2$	+0.000 000 337 2
$weak_1$	$-\alpha^2/4\pi$	-0.000 004 237 6
$weak_2$	$+6\alpha^3/4\pi$	+0.000 000 185 5
$weak_3$	$-6(32\alpha)^1\alpha^3/4\pi$	-0.000 000 043 6
$weak_4$	$+6(32\alpha)^3\alpha^3/4\pi$	+0.000 000 002 3
$weak_5$	$-6(32\alpha)^4\alpha^3/4\pi$	-0.000 000 000 5
$weak_6$	$+6(32\alpha)^5\alpha^3/4\pi$	+0.000 000 000 1
$weak_7$	$-6(32\alpha)^6\alpha^3/4\pi$	-0.000 000 000 0
$weak_8$	$+6(32\alpha)^7\alpha^3/4\pi$	+0.000 000 000 0
$weak_9$	$-6(32\alpha)^8\alpha^3/4\pi$	-0.000 000 000 0
$weak_{10}$	$+6(32\alpha)^9\alpha^3/4\pi$	+0.000 000 000 0

total calculated g/2 factor for $-e_2$ +0.863 605 799 0

Table 5-9
Positron family
$-e_3$ g/2 Factor Evaluation with 2010 α

force:	g/2 factor term:	numerical value:
strong	$+1$	+1.000 000 000 000
meso-electric	$-bn\pi\alpha$	-0.412 655 679 116
electric	$+\alpha/2\pi$	+0.001 161 409 733
magnetic	$+\alpha^2/16\pi^2$	+0.000 000 337 218
$weak_1$	$+\alpha^2/4\pi$	+0.000 004 237 608
$weak_2$	$-10\alpha^3/4\pi$	-0.000 000 309 233
$weak_3$	$+10(32\alpha)^1\alpha^3/4\pi$	+0.000 000 072 210
$weak_4$	$-10(32\alpha)^3\alpha^3/4\pi$	-0.000 000 005 937
$weak_5$	$+10(32\alpha)^4\alpha^3/4\pi$	+0.000 000 001 919
$weak_6$	$-10(32\alpha)^5\alpha^3/4\pi$	-0.000 000 000 214
$weak_7$	$+10(32\alpha)^6\alpha^3/4\pi$	+0.000 000 000 050
$weak_8$	$-10(32\alpha)^7\alpha^3/4\pi$	-0.000 000 000 011
$weak_9$	$+10(32\alpha)^8\alpha^3/4\pi$	+0.000 000 000 002
$weak_{10}$	$-10(32\alpha)^9\alpha^3/4\pi$	-0.000 000 000 000
$weak_{11}$	$+10(32\alpha)^{10}\alpha^3/4\pi$	+0.000 000 000 000

total calculated g/2 factor for $-e_3$ +0.588 510 166 228

Table 5-10
Positron family
-e_4 g/2 Factor Evaluation with 2010 α

force:	g/2 factor term:	numerical value:
strong	$+1$	+1.000 000 000 000
meso-electric	$-bn\pi\alpha$	-0.917 012 368 931
electric	$+\alpha / 2\pi$	+0.001 161 409 733
magnetic	$+\alpha^2 / 16\pi^2$	+0.000 000 337 218
$weak_1$	$-\alpha^2 / 4\pi$	-0.000 004 237 608
$weak_2$	$+15\alpha^3 / 4\pi$	+0.000 000 463 849
$weak_3$	$-15(32\alpha)^1 \alpha^3 / 4\pi$	-0.000 000 103 316
$weak_4$	$+15(32\alpha)^3 \alpha^3 / 4\pi$	+0.000 000 005 633
$weak_5$	$-15(32\alpha)^4 \alpha^3 / 4\pi$	-0.000 000 001 315
$weak_6$	$+15(32\alpha)^5 \alpha^3 / 4\pi$	+0.000 000 000 307
$weak_7$	$-15(32\alpha)^6 \alpha^3 / 4\pi$	-0.000 000 000 071
$weak_8$	$+15(32\alpha)^7 \alpha^3 / 4\pi$	+0.000 000 000 016
$weak_9$	$-15(32\alpha)^8 \alpha^3 / 4\pi$	-0.000 000 000 003
$weak_{10}$	$+15(32\alpha)^9 \alpha^3 / 4\pi$	+0.000 000 000 000
$weak_{11}$	$-15(32\alpha)^{10} \alpha^3 / 4\pi$	-0.000 000 000 000

total calculated g/2 factor for -e_4 +0.084 145 505 542

Table 5-11
Pion family
Pion π_1 g/2 Factor Evaluation with 2010 α

force:	g/2 factor term:	numerical value:
strong	$+1$	+1.000 000 000 000
meso-electric	$-(b-1)n\pi\alpha$	-0.000 000 000 000
electric	$+\alpha/2\pi$	+0.001 161 409 733
magnetic	$+\alpha^2/16\pi^2$	+0.000 000 337 218
$weak_1$	$-\alpha^2/4\pi$	-0.000 004 237 608
$weak_2$	$+\alpha^3/4\pi$	+0.000 000 030 923
$weak_3$	$-(32\alpha)^1\alpha^3/4\pi$	-0.000 000 007 221
$weak_4$	$+(32\alpha)^3\alpha^3/4\pi$	+0.000 000 000 393
$weak_5$	$-(32\alpha)^4\alpha^3/4\pi$	-0.000 000 000 091
$weak_6$	$+(32\alpha)^5\alpha^3/4\pi$	+0.000 000 000 021
$weak_7$	$-(32\alpha)^6\alpha^3/4\pi$	-0.000 000 000 005
$weak_8$	$+(32\alpha)^7\alpha^3/4\pi$	+0.000 000 000 001
$weak_9$	$-(32\alpha)^8\alpha^3/4\pi$	-0.000 000 000 000

total calculated g/2 factor for pion   +1.001 157 533 365

Table 5-12
Pion family
Kaon π_2 g/2 Factor Evaluation with 2010 α

force:	g/2 factor term:	numerical value:
strong	$+1$	+1.000 000 000 000
meso-electric	$-(b-1)n\pi\alpha$	-0.022 925 309 223

electric	$+\alpha/2\pi$	+0.001 161 409 733
magnetic	$+\alpha^2/16\pi^2$	+0.000 000 337 218
weak$_1$	$+\alpha^2/4\pi$	+0.000 004 237 608
weak$_2$	$-3\alpha^3/4\pi$	-0.000 000 092 769
weak$_3$	$+3(32\alpha)^1\,\alpha^3/4\pi$	+0.000 000 021 663
weak$_4$	$-3(32\alpha)^3\,\alpha^3/4\pi$	-0.000 000 001 181
weak$_5$	$+3(32\alpha)^4\,\alpha^3/4\pi$	+0.000 000 000 275
weak$_6$	$-3(32\alpha)^5\,\alpha^3/4\pi$	-0.000 000 000 064
weak$_7$	$+3(32\alpha)^6\,\alpha^3/4\pi$	+0.000 000 000 015
weak$_8$	$-3(32\alpha)^7\,\alpha^3/4\pi$	-0.000 000 000 003
weak$_9$	$+3(32\alpha)^8\,\alpha^3/4\pi$	+0.000 000 000 000
weak$_{10}$	$-3(32\alpha)^9\,\alpha^3/4\pi$	-0.000 000 000 000

total calculated g/2 factor for π_2 +0.978 240 603 272

Table 5-13
Pion family
D-on π_3 g/2 Factor Evaluation with 2010 α

force:	g/2 factor term:	numerical value:
strong	$+1$	$+1.000\ 000\ 000\ 000$
meso-electric	$-(b-1)n\pi\alpha$	$-0.137\ 551\ 874\ 189$
electric	$+\alpha/2\pi$	$+0.001\ 161\ 409\ 727$
magnetic	$+\alpha^2/16\pi^2$	$+0.000\ 000\ 337\ 218$
$weak_1$	$-\alpha^2/4\pi$	$-0..000\ 004\ 237\ 608$
$weak_2$	$+6\alpha^3/4\pi$	$+0.000\ 000\ 185\ 539$
$weak_3$	$-6(32\alpha)^1\,\alpha^3/4\pi$	$-0.000\ 000\ 043\ 663$
$weak_4$	$+6(32\alpha)^3\,\alpha^3/4\pi$	$+0.000\ 000\ 002\ 362$
$weak_5$	$-6(32\alpha)^4\,\alpha^3/4\pi$	$-0.000\ 000\ 000\ 551$
$weak_6$	$+6(32\alpha)^5\,\alpha^3/4\pi$	$+0.000\ 000\ 000\ 128$
$weak_7$	$-6(32\alpha)^6\,\alpha^3/4\pi$	$-0.000\ 000\ 000\ 030$
$weak_8$	$+6(32\alpha)^7\,\alpha^3/4\pi$	$+0.000\ 000\ 000\ 007$
$weak_9$	$-6(32\alpha)^8\,\alpha^3/4\pi$	$-0.000\ 000\ 000\ 001$
$weak_{10}$	$+6(32\alpha)^9\,\alpha^3/4\pi$	$+0.000\ 000\ 000\ 000$

total calculated g/2 factor for π_3 $+0.863\ 605\ 778\ 946$

Table 5-14
Pion family
π_4 g/2 Factor Evaluation with 2010 α

force:	g/2 factor term:	numerical value:
strong	$+1$	+1.000 000 000 000
meso-electric	$-(b-1)n\pi\alpha$	-0.412 655 566 198
electric	$+\alpha/2\pi$	+0.001 161 409 733
magnetic	$+\alpha^2/16\pi^2$	+0.000 000 337 218
$weak_1$	$+\alpha^2/4\pi$	+0.000 004 237 608
$weak_2$	$-10\alpha^3/4\pi$	-0.000 000 309 233
$weak_3$	$+10(32\alpha)^1\alpha^3/4\pi$	+0.000 000 072 210
$weak_4$	$-10(32\alpha)^3\alpha^3/4\pi$	-0.000 000 005 937
$weak_5$	$+10(32\alpha)^4\alpha^3/4\pi$	+0.000 000 001 919
$weak_6$	$-10(32\alpha)^5\alpha^3/4\pi$	-0.000 000 000 214
$weak_7$	$+10(32\alpha)^6\alpha^3/4\pi$	+0.000 000 000 050
$weak_8$	$-10(32\alpha)^7\alpha^3/4\pi$	-0.000 000 000 011
$weak_9$	$+10(32\alpha)^8\alpha^3/4\pi$	+0.000 000 000 002
$weak_{10}$	$-10(32\alpha)^9\alpha^3/4\pi$	-0.000 000 000 000
$weak_{11}$	$+10(32\alpha)^{10}\alpha^3/4\pi$	+0.000 000 000 000

total calculated g/2 factor for π_4 +0.588 510 179 146

Table 5-15
Pion family
π_5 g/2 Factor Evaluation with 2010 α

force:	g/2 factor term:	numerical value:
strong	$+1$	$+1.000\ 000\ 000\ 000$
meso-electric	$-(b-1)n\pi\alpha$	$-0.917\ 012\ 368\ 931$
electric	$+\alpha/2\pi$	$+0.001\ 161\ 409\ 733$
magnetic	$+\alpha^2/16\pi^2$	$+0.000\ 000\ 337\ 218$
$weak_1$	$-\alpha^2/4\pi$	$-0.000\ 004\ 237\ 608$
$weak_2$	$+15\alpha^3/4\pi$	$+0.000\ 000\ 463\ 849$
$weak_3$	$-15(32\alpha)^1\alpha^3/4\pi$	$-0.000\ 000\ 103\ 316$
$weak_4$	$+15(32\alpha)^3\alpha^3/4\pi$	$+0.000\ 000\ 005\ 633$
$weak_5$	$-15(32\alpha)^4\alpha^3/4\pi$	$-0.000\ 000\ 001\ 315$
$weak_6$	$+15(32\alpha)^5\alpha^3/4\pi$	$+0.000\ 000\ 000\ 307$
$weak_7$	$-15(32\alpha)^6\alpha^3/4\pi$	$-0.000\ 000\ 000\ 071$
$weak_8$	$+15(32\alpha)^7\alpha^3/4\pi$	$+0.000\ 000\ 000\ 016$
$weak_9$	$-15(32\alpha)^8\alpha^3/4\pi$	$-0.000\ 000\ 000\ 003$
$weak_{10}$	$+15(32\alpha)^9\alpha^3/4\pi$	$+0.000\ 000\ 000\ 000$
$weak_{11}$	$-15(32\alpha)^{10}\alpha^3/4\pi$	$-0.000\ 000\ 000\ 000$

total calculated g/2 factor for π_5 $+0.084\ 145\ 505\ 512$

Table 5-16
Anti-pion family
Anti-pion $-\pi_1$ g/2 Factor Evaluation with 20010 α

force:	g/2 factor term:	numerical value:
strong	-1	-1.000 000 000 000
electric	$-\alpha/2\pi$	-0.001 161 409 733
magnetic	$-\alpha^2/16\pi^2$	-0.000 000 337 218
$weak_1$	$+\alpha^2/4\pi$	+0.000 004 237 608
$weak_2$	$-\alpha^3/4\pi$	-0.000 000 030 923
$weak_3$	$+(32\alpha)^1\alpha^3/4\pi$	+0.000 000 007 221
$weak_4$	$-(32\alpha)^3\alpha^3/4\pi$	-0.000 000 000 393
$weak_5$	$+(32\alpha)^4\alpha^3/4\pi$	+0.000 000 000 091
$weak_6$	$-(32\alpha)^5\alpha^3/4\pi$	-0.000 000 000 021
$weak_7$	$+(32\alpha)^6\alpha^3/4\pi$	+0.000 000 000 005
$weak_8$	$-(32\alpha)^7\alpha^3/4\pi$	-0.000 000 000 001
$weak_9$	$+(32\alpha)^8\alpha^3/4\pi$	+0.000 000 000 000

total calculated g/2 factor for $-\pi_1$ -1.001 157 533 365

Table 5-17
Anti-pion family
Anti-kaon $-\pi_2$ g/2 Factor Evaluation with 2010 α

force:	g/2 factor term:	numerical value:
strong	-1	-1.000 000 000 000
electric	$-\alpha/2\pi$	-0.001 161 409 733
magnetic	$-\alpha^2/16\pi^2$	-0.000 000 337 218
$weak_1$	$-\alpha^2/4\pi$	-0.000 004 237 608
$weak_2$	$+3\alpha^3/4\pi$	+0.000 000 092 769
$weak_3$	$-3(32\alpha)^1\alpha^3/4\pi$	-0.000 000 021 663
$weak_4$	$+3(32\alpha)^3\alpha^3/4\pi$	+0.000 000 001 181
$weak_5$	$-3(32\alpha)^4\alpha^3/4\pi$	-0.000 000 000 275
$weak_6$	$+3(32\alpha)^5\alpha^3/4\pi$	+0.000 000 000 064
$weak_7$	$-3(32\alpha)^6\alpha^3/4\pi$	-0.000 000 000 015
$weak_8$	$+3(32\alpha)^7\alpha^3/4\pi$	+0.000 000 000 003
$weak_9$	$-3(32\alpha)^8\alpha^3/4\pi$	-0.000 000 000 000
$weak_{10}$	$+3(32\alpha)^9\alpha^3/4\pi$	+0.000 000 000 000

total calculated g/2 factor for $-\pi_2$ -1.001 165 912 495

Table 5-18
Anti-pion family
Anti-D-on $-\pi_3$ g/2 Factor Evaluation with 2010 α

force:	g/2 factor term:	numerical value:
strong	-1	$-1.000\ 000\ 000\ 000$
electric	$-\alpha/2\pi$	$-0.001\ 161\ 409\ 733$
magnetic	$-\alpha^2/16\pi^2$	$-0.000\ 000\ 337\ 218$
$weak_1$	$+\alpha^2/4\pi$	$+0.000\ 004\ 237\ 608$
$weak_2$	$-6\alpha^3/4\pi$	$-0.000\ 000\ 185\ 539$
$weak_3$	$+6(32\alpha)^1\,\alpha^3/4\pi$	$+0.000\ 000\ 043\ 663$
$weak_4$	$-6(32\alpha)^3\,\alpha^3/4\pi$	$-0.000\ 000\ 002\ 362$
$weak_5$	$+6(32\alpha)^4\,\alpha^3/4\pi$	$+0.000\ 000\ 000\ 551$
$weak_6$	$-6(32\alpha)^5\,\alpha^3/4\pi$	$-0.000\ 000\ 000\ 128$
$weak_7$	$+6(32\alpha)^6\,\alpha^3/4\pi$	$+0.000\ 000\ 000\ 030$
$weak_8$	$-6(32\alpha)^7\,\alpha^3/4\pi$	$-0.000\ 000\ 000\ 007$
$weak_9$	$+6(32\alpha)^8\,\alpha^3/4\pi$	$+0.000\ 000\ 000\ 001$
$weak_{10}$	$-6(32\alpha)^7\,\alpha^3/4\pi$	$-0.000\ 000\ 000\ 000$

total calculated g/2 factor for $-\pi_3$ $-1.001\ 157\ 653\ 136$

Table 5-19
Anti-pion family
$-\pi_4$ g/2 Factor Evaluation with 2010 α

force:	g/2 factor term:	numerical value:
strong	-1	-1.000 000 000 000
electric	$-\alpha/2\pi$	-0.001 161 409 733
magnetic	$-\alpha^2/16\pi^2$	-0.000 000 337 218
$weak_1$	$-\alpha^2/4\pi$	-0.000 004 237 608
$weak_2$	$+10\alpha^3/4\pi$	+0.000 000 309 233
$weak_3$	$-10(32\alpha)^1\,\alpha^3/4\pi$	-0.000 000 072 210
$weak_4$	$+10(32\alpha)^3\,\alpha^3/4\pi$	+0.000 000 005 937
$weak_5$	$-10(32\alpha)^4\,\alpha^3/4\pi$	-0.000 000 001 919
$weak_6$	$+10(32\alpha)^5\,\alpha^3/4\pi$	+0.000 000 000 214
$weak_7$	$-10(32\alpha)^6\,\alpha^3/4\pi$	-0.000 000 000 050
$weak_8$	$+10(32\alpha)^7\,\alpha^3/4\pi$	+0.000 000 000 011
$weak_9$	$-10(32\alpha)^8\,\alpha^3/4\pi$	-0.000 000 000 002
$weak_{10}$	$+10(32\alpha)^9\,\alpha^3/4\pi$	+0.000 000 000 000
$weak_{11}$	$-10(32\alpha)^{10}\,\alpha^3/4\pi$	-0.000 000 000 000

total calculated g/2 factor for $-\pi_4$ -1.001 165 744 345

Table 5-20
Anti-pion family
$-\pi_5$ g/2 Factor Evaluation with 20010 α

force:	g/2 factor term:	numerical value:
strong	-1	-1.000 000 000 000
electric	$-\alpha/2\pi$	-0.001 161 409 733
magnetic	$-\alpha^2/16\pi^2$	-0.000 000 337 218
$weak_1$	$+\alpha^2/4\pi$	+ 0.000 004 237 608
$weak_2$	$-15\alpha^3/4\pi$	-0.000 000 463 849
$weak_3$	$+15(32\alpha)^1\,\alpha^3/4\pi$	+0.000 000 103 316
$weak_4$	$-15(32\alpha)^3\,\alpha^3/4\pi$	-0.000 000 005 633
$weak_5$	$+15(32\alpha)^4\,\alpha^3/4\pi$	+0.000 000 001 315
$weak_6$	$-15(32\alpha)^5\,\alpha^3/4\pi$	-0.000 000 000 307
$weak_7$	$+15(32\alpha)^6\,\alpha^3/4\pi$	+0.000 000 000 071
$weak_8$	$-15(32\alpha)^7\,\alpha^3/4\pi$	-0.000 000 000 016
$weak_9$	$+15(32\alpha)^8\,\alpha^3/4\pi$	+0.000 000 000 003
$weak_{10}$	$-15(32\alpha)^9\,\alpha^3/4\pi$	-0.000 000 000 000
$weak_{11}$	$+15(32\alpha)^{10}\,\alpha^3/4\pi$	+0.000 000 000 000

total calculated g/2 factor for $-\pi_5$ -1.001 157 869 444

Table 5-21
Neutron family
Neutron n_1 g/2 Factor Evaluation with 2010 α

force:	g/2 factor term:	numerical value:
strong	-1	-1.000 000 000 000
meso-electric	$-bn\pi\alpha$	-0.022 925 309 222
electric	$-\alpha/2\pi$	-0.001 161 409 733
magnetic	$-\alpha^2/16\pi^2$	-0.000 000 337 218
$weak_1$	$+\alpha^2/8\pi$	+0.000 002 118 804
$weak_2$	$-3\alpha^3/4\pi$	-0.000 000 092 770
$weak_3$	$+3(32\alpha)^1 \alpha^3/4\pi$	+0.000 000 021 831
$weak_4$	$-3(32\alpha)^3 \alpha^3/4\pi$	-0.000 000 001 181
$weak_5$	$+3(32\alpha)^4 \alpha^3/4\pi$	+0.000 000 000 275
$weak_6$	$-3(32\alpha)^5 \alpha^3/4\pi$	-0.000 000 000 064
$weak_7$	$+3(32\alpha)^6 \alpha^3/4\pi$	+0.000 000 000 015
$weak_8$	$-3(32\alpha)^7 \alpha^3/4\pi$	-0.000 000 000 003
$weak_9$	$+3(32\alpha)^8 \alpha^3/4\pi$	+0.000 000 000 000
$weak_{10}$	$-3(32\alpha)^9 \alpha^3/4\pi$	-0.000 000 000 000

total calculated g/2 factor for neutron -1.024 085 009 266

Table 5-22
Neutron family
Λ n_2 g/2 Factor Evaluation with 2010 α

force:	g/2 factor term:	numerical value:
strong	-1	-1.000 000 000 000
meso-electric	$-bn\pi\alpha$	-0.137 551 855 339
electric	$-\alpha/2\pi$	-0.001 161 409 733
magnetic	$-\alpha^2/16\pi^2$	-0.000 000 337 218
$weak_1$	$-\alpha^2/4\pi$	-0.000 004 237 608
$weak_2$	$+6\alpha^3/4\pi$	+0.000 000 185 539
$weak_3$	$-6(32\alpha)^1\alpha^3/4\pi$	-0.000 000 043 663
$weak_4$	$+6(32\alpha)^3\alpha^3/4\pi$	+0.000 000 002 362
$weak_5$	$-6(32\alpha)^4\alpha^3/4\pi$	-0.000 000 000 551
$weak_6$	$+6(32\alpha)^5\alpha^3/4\pi$	+0.000 000 000 128
$weak_7$	$-6(32\alpha)^6\alpha^3/4\pi$	-0.000 000 000 030
$weak_8$	$+6(32\alpha)^7\alpha^3/4\pi$	+0.000 000 000 007
$weak_9$	$-6(32\alpha)^8\alpha^3/4\pi$	-0.000 000 000 001
$weak_{10}$	$+6(32\alpha)^9\alpha^3/4\pi$	+0.000 000 000 000
total calculated g/2 factor for n_2		-1.138 716 794 286

Table 5-23
Neutron family
Σ^0 n_3 $_g/2$ Factor Evaluation with 2010 α

force:	g/2 factor term:	numerical value:
strong	-1	-1.000 000 000 000
meso-electric	$-bn\pi\alpha$	-0.412 655 565 996
electric	$-\alpha/2\pi$	-0.001 161 409 733
magnetic	$-\alpha^2/16\pi^2$	-0.000 000 337 218
$weak_1$	$+\alpha^2/4\pi$	+0.000 004 237 608
$weak_2$	$-10\alpha^3/4\pi$	-0.000 000 309 233
$weak_3$	$+10(32\alpha)^1\alpha^3/4\pi$	+0.000 000 072 210
$weak_4$	$-10(32\alpha)^3\alpha^3/4\pi$	-0.000 000 005 937
$weak_5$	$+10(32\alpha)^4\alpha^3/4\pi$	+0.000 000 001 919
$weak_6$	$-10(32\alpha)^5\alpha^3/4\pi$	-0.000 000 000 214
$weak_7$	$+10(32\alpha)^6\alpha^3/4\pi$	+0.000 000 000 050
$weak_8$	$-10(32\alpha)^7\alpha^3/4\pi$	-0.000 000 000 011
$weak_9$	$+10(32\alpha)^8\alpha^3/4\pi$	+0.000 000 000 002
$weak_{10}$	$-10(32\alpha)^9\alpha^3/4\pi$	-0.000 000 000 000
$weak_{11}$	$+10(32\alpha)^{10}\alpha^3/4\pi$	+0.000 000 000 000

total calculated g/2 factor for n_3 -1.413 813 308 461

Table 5-24
Neutron family
$\Lambda_b{}^0$ n_4 g/2 Factor Evaluation with 2010 α

force:	g/2 factor term:	numerical value:
strong	-1	-1.000 000 000 000
meso-electric	$-bn\pi\alpha$	-0.917 012 368 931
electric	$-\alpha / 2\pi$	-0.001 161 409 733
magnetic	$-\alpha^2 / 16\pi^2$	-0.000 000 337 218
$weak_1$	$-\alpha^2 / 4\pi$	-0.000 004 237 608
$weak_2$	$+15\alpha^3 / 4\pi$	+0.000 000 463 849
$weak_3$	$-15(32\alpha)^1 \alpha^3 / 4\pi$	-0.000 000 103 316
$weak_4$	$+15(32\alpha)^3 \alpha^3 / 4\pi$	+0.000 000 005 633
$weak_5$	$-15(32\alpha)^4 \alpha^3 / 4\pi$	-0.000 000 001 315
$weak_6$	$+15(32\alpha)^5 \alpha^3 / 4\pi$	+0.000 000 000 307
$weak_7$	$-15(32\alpha)^6 \alpha^3 / 4\pi$	-0.000 000 000 071
$weak_8$	$+15(32\alpha)^7 \alpha^3 / 4\pi$	+0.000 000 000 016
$weak_9$	$-15(32\alpha)^8 \alpha^3 / 4\pi$	-0.000 000 000 003
$weak_{10}$	$+15(32\alpha)^9 \alpha^3 / 4\pi$	+0.000 000 000 000
$weak_{11}$	$-15(32\alpha)^{10} \alpha^3 / 4\pi$	-0.000 000 000 000

total calculated g/2 factor for n_4 -1.918 178 088 390

Table 5-25
Anti-neutron family
Anti-neutron $-n_1$ g/2 Factor Evaluation with 2010 α

force:	g/2 factor term:	numerical value:
strong	$+1$	+1.000 000 000 000
electric	$+\alpha/2\pi$	+0.001 161 409 733
magnetic	$+\alpha^2/16\pi^2$	+0.000 000 337 218
$weak_1$	$-\alpha^2/4\pi$	-0.000 004 237 608
$weak_2$	$+3\alpha^3/4\pi$	+0.000 000 092 769
$weak_3$	$-3(32\alpha)^1\alpha^3/4\pi$	-0.000 000 021 663
$weak_4$	$+3(32\alpha)^3\alpha^3/4\pi$	+0.000 000 001 181
$weak_5$	$-3(32\alpha)^4\alpha^3/4\pi$	-0.000 000 000 275
$weak_6$	$+3(32\alpha)^5\alpha^3/4\pi$	+0.000 000 000 064
$weak_7$	$-3(32\alpha)^6\alpha^3/4\pi$	-0.000 000 000 015
$weak_8$	$+3(32\alpha)^7\alpha^3/4\pi$	+0.000 000 000 003
$weak_9$	$-3(32\alpha)^8\alpha^3/4\pi$	-0.000 000 000 000
$weak_{10}$	$+3(32\alpha)^9\alpha^3/4\pi$	+0.000 000 000 000

total calculated g/2 factor for $-n_1$ +1.001 157 581 408

Table 5-26
Anti-neutron family
$-\Lambda$ $-n_2$ g/2 Factor Evaluation with 2010 α

force:	g/2 factor term:	numerical value:
strong	$+1$	$+1.000\ 000\ 000\ 000$
electric	$+\alpha/2\pi$	$+0.001\ 161\ 409\ 733$
magnetic	$+\alpha^2/16\pi^2$	$+0.000\ 000\ 337\ 218$
$weak_1$	$+\alpha^2/4\pi$	$+0.000\ 004\ 237\ 608$
$weak_2$	$-6\alpha^3/4\pi$	$-0.000\ 000\ 185\ 539$
$weak_3$	$+6(32\alpha)^1\alpha^3/4\pi$	$+0.000\ 000\ 043\ 663$
$weak_4$	$-6(32\alpha)^3\alpha^3/4\pi$	$-0.000\ 000\ 002\ 362$
$weak_5$	$+6(32\alpha)^4\alpha^3/4\pi$	$+0.000\ 000\ 000\ 551$
$weak_6$	$-6(32\alpha)^5\alpha^3/4\pi$	$-0.000\ 000\ 000\ 128$
$weak_7$	$+6(32\alpha)^6\alpha^3/4\pi$	$+0.000\ 000\ 000\ 030$
$weak_8$	$-6(32\alpha)^7\alpha^3/4\pi$	$-0.000\ 000\ 000\ 007$
$weak_9$	$+6(32\alpha)^8\alpha^3/4\pi$	$+0.000\ 000\ 000\ 001$
$weak_{10}$	$-6(32\alpha)^7\alpha^3/4\pi$	$-0.000\ 000\ 000\ 000$

total calculated g/2 factor for $-n_2$ $+1.001\ 165\ 840\ 767$

Table 5-27
Anti-neutron family
$-\Sigma^0$ $-n_3$ g/2 Factor Evaluation with 2010 α

force:	g/2 factor term:	numerical value:
strong	$+1$	$+1.000\ 000\ 000\ 000$
electric	$+\alpha/2\pi$	$+0.001\ 161\ 409\ 733$
magnetic	$+\alpha^2/16\pi^2$	$+0.000\ 000\ 337\ 218$
$weak_1$	$-\alpha^2/4\pi$	$-0.000\ 004\ 237\ 608$
$weak_2$	$+10\alpha^3/4\pi$	$+0.000\ 000\ 309\ 233$
$weak_3$	$-10(32\alpha)^1\alpha^3/4\pi$	$-0.000\ 000\ 072\ 210$
$weak_4$	$+10(32\alpha)^3\alpha^3/4\pi$	$+0.000\ 000\ 005\ 937$
$weak_5$	$-10(32\alpha)^4\alpha^3/4\pi$	$-0.000\ 000\ 001\ 919$
$weak_6$	$+10(32\alpha)^5\alpha^3/4\pi$	$+0.000\ 000\ 000\ 214$
$weak_7$	$-10(32\alpha)^6\alpha^3/4\pi$	$-0.000\ 000\ 000\ 050$
$weak_8$	$+10(32\alpha)^7\alpha^3/4\pi$	$+0.000\ 000\ 000\ 011$
$weak_9$	$-10(32\alpha)^8\alpha^3/4\pi$	$-0.000\ 000\ 000\ 002$
$weak_{10}$	$+10(32\alpha)^9\alpha^3/4\pi$	$+0.000\ 000\ 000\ 000$
$weak_{11}$	$-10(32\alpha)^{10}\alpha^3/4\pi$	$-0.000\ 000\ 000\ 000$

total calculated g/2 factor for $-n_3$ $+1.001\ 157\ 750\ 558$

Table 5-28
Anti-neutron family
$-\Lambda_b^{\,0}$ $-n_4$ g/2 Factor Evaluation with 2010 α

force:	g/2 factor term:	numerical value:
strong	$+1$	$+1.000\ 000\ 000\ 000$
electric	$+\alpha\,/\,2\pi$	$+0.001\ 161\ 409\ 733$
magnetic	$+\alpha^2\,/\,16\pi^2$	$+0.000\ 000\ 337\ 218$
$weak_1$	$+\alpha^2\,/\,4\pi$	$+\ 0.000\ 004\ 237\ 608$
$weak_2$	$-15\alpha^3\,/\,4\pi$	$-0.000\ 000\ 463\ 849$
$weak_3$	$+15(32\alpha)^1\,\alpha^3\,/\,4\pi$	$+0.000\ 000\ 103\ 316$
$weak_4$	$-15(32\alpha)^3\,\alpha^3\,/\,4\pi$	$-0.000\ 000\ 005\ 633$
$weak_5$	$+15(32\alpha)^4\,\alpha^3\,/\,4\pi$	$+0.000\ 000\ 001\ 315$
$weak_6$	$-15(32\alpha)^5\,\alpha^3\,/\,4\pi$	$-0.000\ 000\ 000\ 307$
$weak_7$	$+15(32\alpha)^6\,\alpha^3\,/\,4\pi$	$+0.000\ 000\ 000\ 071$
$weak_8$	$-15(32\alpha)^7\,\alpha^3\,/\,4\pi$	$-0.000\ 000\ 000\ 016$
$weak_9$	$+15(32\alpha)^8\,\alpha^3\,/\,4\pi$	$+0.000\ 000\ 000\ 003$
$weak_{10}$	$-15(32\alpha)^9\,\alpha^3\,/\,4\pi$	$-0.000\ 000\ 000\ 000$
$weak_{11}$	$+15(32\alpha)^{10}\,\alpha^3\,/\,4\pi$	$+0.000\ 000\ 000\ 000$

total calculated g/2 factor for $-n_4$ $+1.001\ 165\ 619\ 459$

UNITON

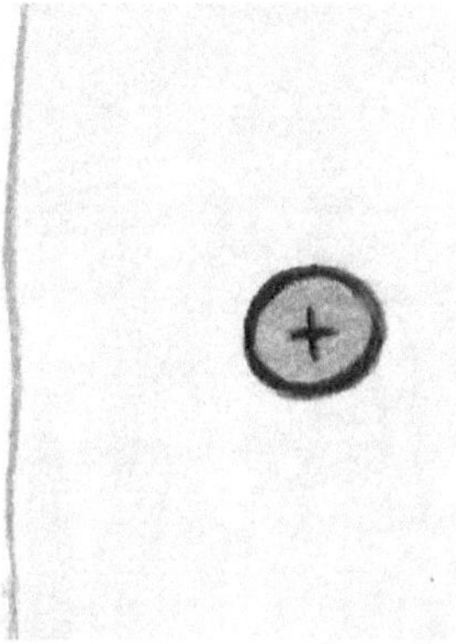

NEUTRON

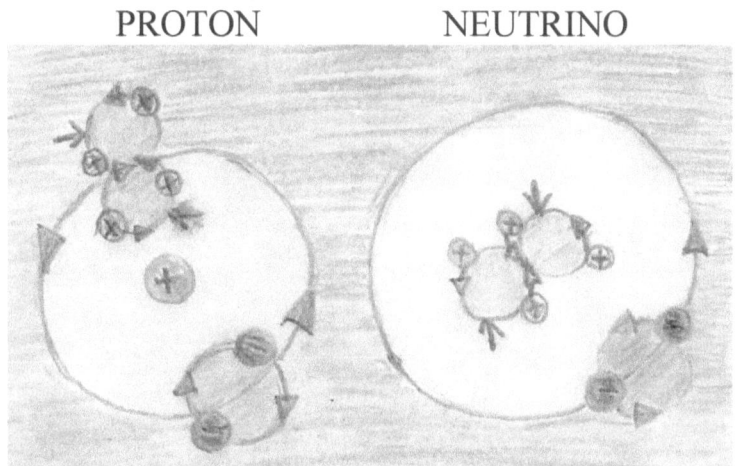

Figure 3. A uniton is a whole electrino. It is the core particle of protons and neutrons and is half of photons. They never come alone.

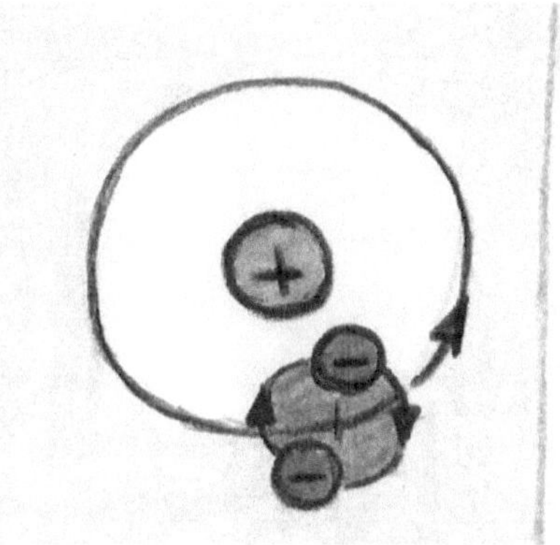

Figure 4. A neutron is a pair of orbiting semions orbiting about a uniton. The total charge is zero.

PROTON

NEUTRINO

Figure 5. A proton is an electron and pion orbiting a uniton.

Figure 6. A neutrino is an an electron orbiting a pion, and traveling near c.

www.ingramcontent.com/pod-product-compliance
Lightning Source LLC
Chambersburg PA
CBHW030913180526
45163CB00004B/1807